In: Intrinsically Disordered Proteins

MEASLES VIRUS NUCLEOPROTEIN

SONIA LONGHI
EDITOR

Nova Biomedical Books
New York

For permission to use material from this book please contact us:
Telephone 631-231-7269; Fax 631-231-8175
Web Site: http://www.novapublishers.com

Library of Congress Cataloging-in-Publication Data

Measles virus nucleoprotein / Sonia Longhi (editor).
p. ; cm.
Includes bibliographical references and index.
ISBN-13: 978-1-60021-629-9 (hardcover)
ISBN-10: 1-60021-629-3 (hardcover)
1. Measles virus. 2. Nucleoproteins. I. Longhi, Sonia.
[DNLM: 1. Measles virus--ultrastructure. 2. Nucleoproteins--physiology. 3. Viral Proteins--physiology. QW 168.5.P2 M4845 2007]
QR201.M43M43 2007
614.5'23072--dc22

2007010438

Published by Nova Science Publishers, Inc. ✦ New York

In: Intrinsically Disordered Proteins

MEASLES VIRUS NUCLEOPROTEIN

INTRINSICALLY DISORDERED PROTEINS

Series Editors: Vladimir N. N. Uversky and A. K. Dunker

Measles Virus Nucleoprotein
Sonia Longhi
2008. ISBN-13: 978-1-60021-629-9 (hardcover)

Contents

Preface

Measles virus possesses a non-segmented, single stranded, negative sense RNA genome that is encapsidated by the nucleoprotein to form a helical nucleocapsid. This ribonucleoproteic complex is the substrate for both transcription and replication. The RNA–dependent RNA polymerase binds to the nucleocapsid template via its co-factor, the phosphoprotein. This book focuses on the main structural information available on the nucleoprotein, showing that it consists of a structured core (NCORE) and of an intrinsically disordered C-terminal domain (NTAIL). The functional implications of the disordered nature of NTAIL are discussed in light of the ability of disordered regions to establish interactions with multiple partners, thus leading to multiple biological effects. Indeed, beyond the phosphoprotein, NTAIL also interacts with cellular partners, including the major heat shock protein, hsp72, the interferon regulator factor 3, IRF3, and a yet unidentified cellular receptor referred to as NR. This book consists of two chapters devoted to the general functions of the nucleoprotein in transcription and replication and to a detailed overview of its structural properties, and of three chapters focused on the functional relevance of the interaction between NTAIL and its various intracellular and extracellular partners.

Chapter 1 - Measles virus belongs to the *Paramyxoviridae* family within the *Mononegavirales* order. Its nonsegmented, single stranded, negative sense RNA genome is encapsidated by the nucleoprotein (N) to form a helical nucleocapsid. This ribonucleoproteic complex is the substrate for both transcription and replication. The RNA–dependent RNA polymerase binds to the nucleocapsid template via its co-factor, the phosphoprotein (P). In this chapter, we describe the main structural information available on the nucleoprotein, showing that it consists of a structured core (N_{CORE}) and of an intrinsically disordered C-terminal domain (N_{TAIL}). We propose a model where the dynamic breaking and reforming of the interaction between N_{TAIL} and P would allow the polymerase complex (L-P) to cartwheel on the nucleocapsid template. We also propose a model where the flexibility of the disordered N and P domains allows the formation of a tripartite complex (N°-P-L) during replication, followed by the delivery of N monomers to the newly synthesized genomic RNA chain. Finally, the functional implications of structural disorder are also discussed in light of the ability of disordered regions to establish interactions with multiple partners, thus leading to multiple biological effects.

Chapter 2 - The intrinsically disordered nature of *Paramyxovirinae* N proteins coupled with the flexibility of N-RNA complexes such as nucleocapsids (NCs), presents significant challenges to their structure analysis. Electron microscopy and image reconstruction has provided a low-resolution insight into NC morphology, highlighting considerable conformational flexibility. NCs range in pitch (the axial rise per helix turn) from 46 to 66 Å and also adopt extended forms with a pitch of 37.5 nm. NCs also vary significantly in twist, with values ranging between at least 12.35 and 13.6 subunits per turn. These variations in conformation appear to be modulated by the natively unfolded C-terminal region, N_{TAIL}, leading to the suggestion that changes in helical parameters may play a role in regulating some aspect of the viral replication cycle, possibly regulating the switch between transcription and replication of the viral genome.

Chapter 3 - Cellular heat shock proteins (HSPs) are induced by numerous physiological stimuli including fever [98, 139]. Historical emphasis was placed upon the function of HSPs as cellular chaperones serving cytoprotective functions [52]. New roles are continually being defined, and we now know that viruses exploit HSP function to support replication in cell culture and that HSPs can significantly modulate both innate and adaptive immune responses [111]. However, the mechanisms by which HSPs enhance gene expression and replication of RNA viruses are only now being elucidated and an understanding of how HSP influences the *in vivo* outcome of virus infection is conspicuously lacking for any viral system. The present chapter will address our understanding of how the major inducible 70 kDa heat shock protein interacts with the C-terminal disordered domain on the measles virus nucleocapsid protein (N_{TAIL}) to influence both viral transcription and genomic replication. Our ability to identify structural determinants of these HSP-mediated changes in viral replication enabled the design of HSP-responsive and non-responsive measles virus variants that, in turn, allowed us to establish the biological significance of HSP-N_{TAIL} structural and functional interactions using a mouse model of measles virus encephalitis.

Chhaper 4 - IFN response, which plays an important part of cellular and immune response to measles virus infection, strongly relies on the activation of IRF-3, a constitutive cytoplasmic transcription factor. The sensing of a "danger" signal by receptors induces the activation of TBK1, which phosphorylates, IRF-3. Then, IRF-3 homo- or heterodimerizes, translocates to the nucleus and, along with NF-κB and AP-1, activates the expression of IFN-α/β and proinflammatory cytokines. Measles virus displays another IRF-3 activation route: the nucleoprotein (N), via its N_{TAIL} region, binds to IRF-3 IAD domain and somehow recruits TBK-1 to phosporylate IRF-3. Interestingly, only a subset of IRF-3-dependent genes (not including IFN-β) is activated through this interaction. The underlying mechanism is yet to be unravelled.

Chapter 5 - The major physiopathological feature of measles virus (MeV) infection is the induction of a specific long lasting anti-viral immune response, which coincides with the appearance of a profound and transient immunosuppression. On one hand, the adaptive immune response efficiency is illustrated by the clearance of the infection within 2 weeks and by the long-life protection against reinfection. On the other hand, the immunosuppression is characterized by a dramatic impairment of humoral and cellular immune responses to unrelated antigens. This feature is induced by direct infection and indirect mechanisms, such as alteration of the biology of uninfected cells by viral proteins. The surprising observation, that MeV nucleoprotein (N) antibodies are the first to be produced in high amounts after MeV infection provides the basis for the hypothesis that the cytosolic MeV N could be released in the extracellular milieu and therefore interact with various extracellular partners with a significant impact on immune responses. This unforeseen role of MeV N involves a dual activity in the development of both MeV-specific immune response and immunosuppression. The analysis of MeV N in both *in vitro* and *in vivo* experimental models, as well as in human patients with measles revealed the complexity of the mechanism of action of this protein.

In: Measles Virus Nucleoprotein
Editors: S. Longhi, pp. 1-36

ISBN: 978-1-60021-629-9

Chapter I

Measles Virus Nucleoprotein: Structural Organization and Functional Role of the Intrinsically Disordered C-Terminal Domain

Jean-Marie Bourhis and Sonia Longhi*
Architecture et Fonction des Macromolécules Biologiques,
UMR 6098 CNRS et Universités Aix-Marseille I et II,
Campus de Luminy, 13288 Marseille Cedex 09, France

Abstract

Measles virus belongs to the *Paramyxoviridae* family within the *Mononegavirales* order. Its nonsegmented, single stranded, negative sense RNA genome is encapsidated by the nucleoprotein (N) to form a helical nucleocapsid. This ribonucleoproteic complex is the substrate for both transcription and replication. The RNA–dependent RNA polymerase binds to the nucleocapsid template via its co-factor, the phosphoprotein (P). In this chapter, we describe the main structural information available on the nucleoprotein, showing that it consists of a structured core (N_{CORE}) and of an intrinsically disordered C-terminal domain (N_{TAIL}). We propose a model where the dynamic breaking and reforming of the interaction between N_{TAIL} and P would allow the polymerase complex (L-P) to cartwheel on the nucleocapsid template. We also propose a model where the flexibility of the disordered N and P domains allows the formation of a tripartite complex (N°-P-L) during replication, followed by the delivery of N monomers to the newly synthesized genomic RNA chain. Finally, the functional implications of structural disorder are also discussed in light of the ability of disordered regions to establish interactions with multiple partners, thus leading to multiple biological effects.

* Corresponding author: CNRS-AFMB Case 932 163, avenue de Luminy, 13288 Marseille Cedex 09, France. E-mail sonia.longhi@afmb.univ-mrs.fr. Tel: (33) 4 91 82 55 80. Fax: (33) 4 91 26 67 20.

List of Abbreviations

CCR	Central Conserved Region
CD	Circular Dichroism
CDV	Canine Distemper Virus
D_{max}	Maximal intramolecular distance
EPR	Electron Paramagnetic Resonance
F	Fusion protein
H	Attachment protein
HN-NMR	Heteronuclear-NMR
hsp72	major inducible 70 kDa heat shock protein
IRF-3	Interferon Regulator Factor 3
IUP	Intrinsically Unstructured Protein
K_D	Equilibrium dissociation constant
L	Large protein
L-P	Polymerase complex
M	Matrix protein
MeV	Measles Virus
MoRE	Molecular Recognition Element
MTSL	Methane Thiosulphonate
N	Nucleoprotein
N°	Monomeric form of N
N_{CORE}	Structured N-terminal domain of N
N^{NUC}	Assembled form of N
N_{TAIL}	Unstructured C-terminal domain of N
NES	Nuclear Export Signal
NLS	Nuclear Localization Signal
NMR	Nuclear Magnetic Resonance
2D-NMR	Bidimensional-NMR
NR	Nucleoprotein Receptor
P	Phosphoprotein
PCT	C-Terminal Domain of P
PDB	Protein Data Bank
PMD	Multimerization Domain of P
PNT	N-Terminal Domain of P
RdRp	RNA-dependent RNA-polymerase
RNA	Ribonucleic acid
RNP	Ribonucleoproteic complex
R_g	Radius of gyration
SAXS	Small Angle X-ray Scattering
SDSL	Site Directed Spin Labeling
TFE	Tri-fluoro-ethanol
VSV	Vesicular Stomatitis Virus
XD	X Domain of P

1. Introduction

Measles virus (MeV) is a highly contagious human pathogen responsible for about half million of deaths in 2004, with about 40 million infections worldwide each year (http://www.who.int/mediacentre/factsheets/fs286/en/). It belongs to the *Paramyxoviridae* family within the *Mononegavirales* order. This order includes several human pathogens with a strong socio-economical impact and comprises both well characterized viruses (as for instance mumps, parainfluenza, rabies and Ebola viruses) and emerging viruses, such as the Nipah and Hendra viruses that are responsible for encephalitis with a high (>50%) case-fatality rate. With approximately 450,000 deaths worldwide, measles ranks 8th as the cause of worldwide mortality and represents the main cause of childhood mortality in developing countries. Despite extensive vaccination campaigns, the disease has not been eradicated yet. Furthermore, outbreaks occur even within vaccinated populations [139, 140]. To date, no effective antiviral treatment exists.

The replicative complex of MeV possesses some features that make it a promising target for the establishment of an antiviral chemotherapy. In particular, the RNA polymerase RNA dependent (RdRp) has no known homologue in human or animal tissues, and it exclusively uses a ribonucleoprotein complex as the substrate for both transcription and replication. Despite their relevance in terms of human health, our understanding of the molecular mechanisms of transcription and replication of *Paramyxoviridae* is still rather poor (see [2, 81, 84] for reviews).

The elucidation of the mechanisms of action of proteins of the replicative complex of MeV, as well as their structural characterization, is a prerequisite for the identification and the rational design of antiviral agents. In this regard, the discovery that the N protein establishes many protein-protein interactions trough its intrinsically disordered C-terminal domain is particularly relevant: indeed protein-protein interactions mediated by disordered regions provide interesting drug discovery targets with the potential to increase significantly the discovery rate for new compounds [28]. These latter could eventually be used against MeV and/or other *Mononegavirales* pathogens.

2. The Virus and Its Replicative Complex

MeV is an enveloped and pleiomorphic virus, within the *Morbillivirus* genus. Its envelope is composed of a lipid bilayer, derived from the plasma membrane of the host cell, which contains the attachment (H) and fusion (F) proteins. Beneath the envelope, the viral matrix protein (M) associates with the cytoplasmic tails of the H and F proteins as well as the viral core particle or nucleocapsid. MeV is a negative-strand RNA virus, i.e. it has a genome of polarity opposed to that of viral mRNAs. Its non-segmented RNA genome is 15 894 nucleotides long and its sequence has been determined [24]. The viral genome is encapsidated by the nucleoprotein (N) to form a helical nucleocapsid. The viral RNA is tightly bound within the nucleocapsid and does not dissociate during RNA synthesis, as shown by its resistance to silencing by siRNA [10]. Therefore, this ribonucleoproteic (RNP) complex, rather than naked RNA, is used as a template for both transcription and replication.

These latter activities are carried out by the RdRp that is composed of the large (L) protein) and of the phosphoprotein (P). The P protein is an essential polymerase co-factor in that it tethers the L protein onto the nucleocapsid template. This ribonucleoproteic complex made of RNA, N, P and L constitutes the replicative unit (Figure 1A).

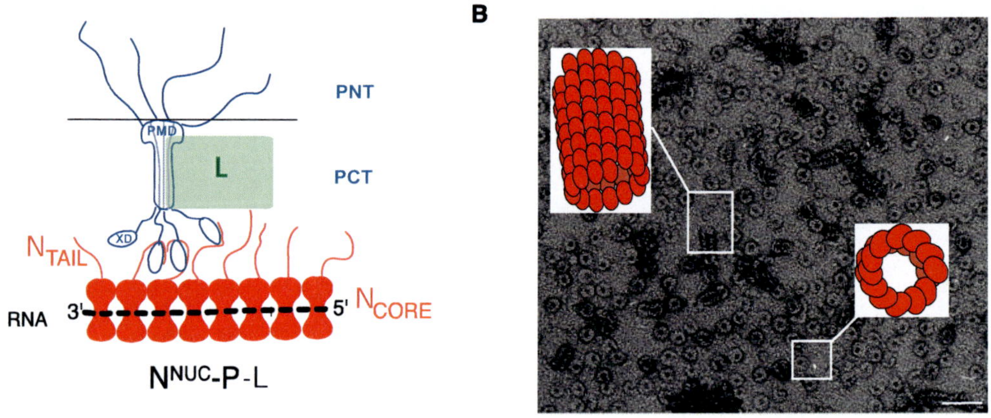

Figure 1. (A) Schematic representation of the N^{NUC}-P-L complex of measles virus. The disordered N_{TAIL} (aa 401-525) and PNT (aa 1-230) regions are represented by lines. The nucleoprotein is represented in red, and the phosphoprotein in blue. The encapsidated RNA is shown as a black dotted line embedded in the middle of N by analogy with *Rhabdoviridae* N-RNA complexes [1, 57]. The multimerization domain of P (aa 304-375, PMD) is represented with a dumbbell shape according to Tarbouriech *et al.* [120]. The tetrameric P [103] is shown bound to N^{NUC} through 3 of its 4 C-terminal XD (aa 459-507) "arms", as in the model of Curran and Kolakofsky [31]. The segment connecting PMD and XD is represented as disordered according to [85] and [70]. The L protein is shown as a rectangle contacting P through PMD by analogy with SeV [114]. (B) Negative-staining electron micrographs of bacterially expressed MeV N. Nucleocapsid-like, herringbone structures, are shown on the left, while rings, corresponding to short nucleocapsids seen perpendicularly to their axis, are shown on the right. The bar represents 50 nm. Data of panel B were taken from [71].

MeV nucleocapsids, as visualized by negative stain transmission electron microscopy, have a typical herringbone-like appearance (Figure 1B) (see also Chapter 2). The nucleocapsid of all *Paramyxoviridae* has a considerable conformational flexibility and can adopt different helical pitches (the axial rise *per* turn) and twistes (the number of subunits per turn) resulting in conformations differing in their extent of compactness [7, 8, 45, 62, 63, 112]. In the case of the MeV nucleocapsid, the most extended conformation has a helical pitch of 66 Å, while twist varies from 13.04 to 13.44 [8] (see also Chapter 2).

Once the viral RNPs are released in the cytoplasm of infected cells, transcription of viral genes occurs using endogenous NTPs as substrates. Each gene is flanked by untranslated 3' and 5' regions. Between the gene-end boundaries are intergenic regions, which contain exactly 3 nucleotides. During transcription, the polymerase skips these intergenic regions and re-initiates synthesis at the downstream promoter. As the replication cycle of *Mononegavirales* (with the exception of the *Bornaviridae* family) takes place in the cytoplasm, these viruses have to synthesize their own mRNA 5'-terminal cap structure. Following primary transcription, the polymerase switches to a processive mode and ignores the gene junctions to synthesize a full, complementary strand of genome length. This positive-stranded RNA (antigenome) does not serve as a template for transcription, and its unique role is to provide an intermediate in replication (Figure 2). The intracellular concentration of the N protein is the main element controlling the relative rates of transcription and replication. When N is limiting, the polymerase is preferentially engaged in transcription, thus leading to an increase in the intracellular concentration of viral proteins, including N. When N levels are high enough to allow encapsidation of the nascent RNA chain, the polymerase switches to a replication mode [101] (see [2, 81, 84, 108] for reviews on transcription and replication).

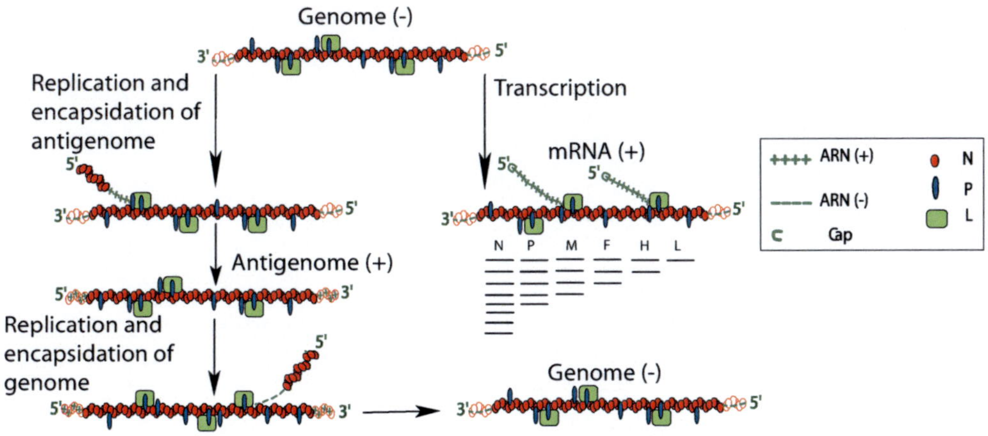

Figure 2. Schematic representation of transcription and replication of MeV genome. Genomic RNA is defined as being of (-) polarity, and viral mRNA of (+) polarity. (+) RNAs and (-) RNAs are represented as stretches of green (+) and (-) signs, respectively. The transcribed viral mRNAs are capped, as symbolized by a "C" at the 5' end of the mRNA. The viral mRNAs have been drawn in different amounts to take into account the effective mRNA gradient in infected cells [101]. During replication, genomic RNA is replicated into a full-length complementary copy, the antigenome, of (+) polarity. The antigenome is encapsidated concomitantly to its synthesis. It is not known however whether the nascent antigenomic chain is encapsidated as soon as it is synthesized (as depicted here for clarity) or with a slight delay. Antigenomic RNA only serves as a template for replication and is thus in turn replicated into a genomic RNA of (-) polarity, which is concurrently encapsidated. The neosynthesized genomic RNA is a template for further mRNA synthesis, called secondary transcription (not shown).

The nucleoprotein is the most abundant structural protein. Its primary function is to encapsidate the viral genome. However, as we will see in this and in the following chapters, N is not a simple structural component, serving merely to package viral RNA. Rather, it plays several functions. Within *Mononegavirales*, each N monomer interacts with a precise number of nucleotides. The number of nucleotides varies amongst *Mononegavirales*, being specific to each family: 6 nucleotides for *Paramyxoviridae* [45], 9 for *Rhabdoviridae* [1, 52, 57], and 12-15 for *Filoviridae* [87]. The fact that in Paramyxoviridae N wraps exactly 6 nucleotides imposes the so-called "rule of six" to these viruses, i.e. their genome must be of polyhexameric length (6n + 0) to efficiently replicate (see [75, 108, 135] for reviews).

The specific encapsidation signal is thought to lay within the 5' *Leader* and *Trailer* extremities of the antigenome and genome RNA strands, respectively, as demonstrated for another *Mononegavirales* member [12]. However, in the absence of viral RNA and of other viral proteins, *Mononegavirales* nucleoproteins are able to self-assemble onto cellular RNA to form nucleocapsid-like particles. Therefore, a regulatory mechanism is necessary to prevent the illegitimate self-assembly of N in the absence of ongoing genomic RNA synthesis. Indeed, in infected cells N is found in various forms: a soluble, monomeric form (referred to as N°) and an assembled form (referred to as N^{NUC}). Two forms of soluble N are likely to occur in the cytosol: a neosynthesized, transient form (herein referred to as n) and a more mature form, N° (Figure 3) [56]. The soluble form of N is localized in both the cytosol and the nucleus (Figure 3) [56, 64]. Sato and co-workers have recently identified the determinants of the intracellular trafficking within the nucleoprotein sequence of MeV and canine distemper virus (CDV), a closely related *Morbillivirus*. They both possess a novel nuclear localization signal (NLS) at positions 70-77 and a nuclear export signal (NES). The NLS has a novel leucine/isoleucine-rich motif (TGILISIL), whereas the NES is composed of a leucine-rich motif (LLRSLTLF). While in CDV the NES occurs at positions 4-11, in MeV it is located in the C-terminus [110]. In both viruses, the nuclear export of N is CRM1-independent.

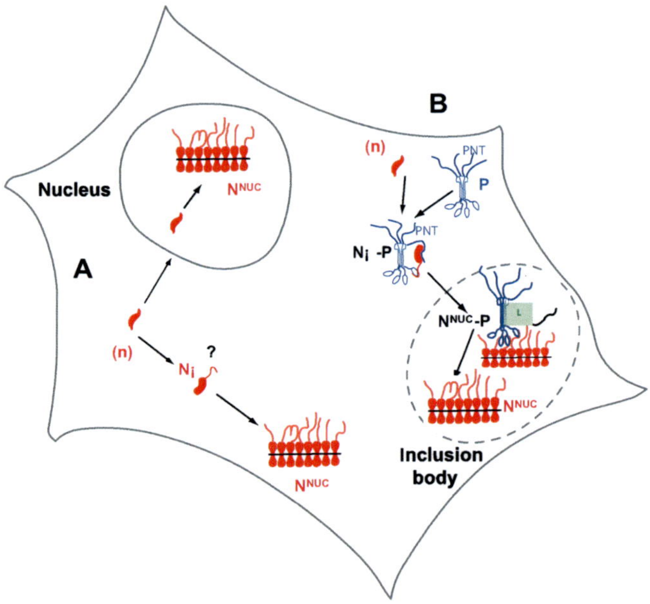

Figure 3. Maturation of MeV N according to [56]. Interior of an infected cell, with the nucleus (top left) and a cytoplasmic inclusion body (bottom right). N is found under several conformations. The neosynthesized form (n) and a more mature form (N°) are both cytosolic, whereas the assembled form (N^{NUC}) is probably bound to the cytoskeleton (not shown). The polymerase complex is formed by L (green) and by a tetramer of P (blue). (A) In the absence of P, n can self-assemble illegitimately on cellular RNA (bottom). It can also migrate to the nucleus, where it forms nucleocapsid-like particles. It is not known whether it undergoes a conformational change directly from n to N^{NUC} or whether N° is an intermediate conformation in the process. (B) P forms a complex with N° thereby preventing illegitimate self-assembly of N. The N°-P complex has been represented with a 1:4 stoichiometry by analogy to SeV (J. Curran, personal communication). Within the N°-P complex, the N_{CORE} region has been represented with a shape slightly differing from that of the N^{NUC}-P complex according to [56]. The N°-P is used by the polymerase to encapsidate neosynthesized RNA during replication (which takes place in the cytoplasmic inclusion bodies).

Once synthesized, the monomeric form of N requires the presence of a chaperone. This role is played by the P protein, whose association with N prevents illegitimate self-assembly of this latter, and also retains the soluble form of N in the cytoplasm [66, 115]. This soluble N°-P complex is used as a substrate for the encapsidation of nascent genomic RNA chain during replication. The assembled form of N also forms complexes with P, either isolated (N^{NUC}-P) or bound to L (N^{NUC}-P-L), which are essential to RNA synthesis by the viral polymerase [21, 109] (Figure 1A and Figure 3).

As the nucleoprotein, the P protein provides several functions in transcription and replication. Beyond serving as a chaperone for N, P binds to the nucleocapsid, thus tethering the polymerase onto the nucleocapsid template. P is a modular protein, consisting of at least two domains: an N-terminal disordered domain (aa 1-230, PNT) [72], and a C-terminal domain (aa 231-507, PCT) (for a more detailed description of the modular organization of P see [16, 17]). Transcription requires only the PCT domain, whereas genome replication also requires PNT.

The viral polymerase, which is responsible for both transcription and replication, is poorly characterized. It is thought to carry out most (if not all) enzymatic activities required for transcription and replication, including nucleotide polymerization, mRNA capping and polyadenylation. However, no *Paramyxoviridae* polymerase has been purified so far, implying that most of our present knowledge arises from bioinformatics studies. Notably, using bioinformatics approaches we identified a ribose-2'-O-methyltransferase domain involved in capping of viral mRNAs within the C-terminal region of *Mononegavirales* polymerases (with the exception of *Bornaviridae* and *Nucleorhabdoviruses*) [49]. The methyltransferase activity of the C-terminal region of Sendai virus (SeV) polymerase (aa 1756-2228) has been recently demonstrated biochemically [94]. Interestingly, the polymerase of vesicular stomatitis virus (VSV), a *Rhabdoviridae* member, has been recently shown to possess both ribose-2'-O and guanine-N-7-methyltransferase activities [82].

In all *Mononegavirales* members, the viral genomic RNA is always encapsidated by the N protein, and genomic replication does not occur in the absence of N°. Therefore, during RNA synthesis the viral polymerase has to interact with the N:RNA complex, and to use the N°-P complex as the substrate of encapsidation. Hence, the components of the viral replication machinery, namely P, N and L, engage in a complex macromolecular ballet. Although the understanding of the roles of N, P and L within the replicative complex of MeV has benefited of significant breakthroughs in recent years (see [17] for a review), rather limited three-dimensional information on the replicative machinery is available. The scarcity of high-resolution structural data stems from several facts: *i*) the difficulty to obtain homogenous polymers of N suitable for X-ray analysis [71, 111], *ii*) the low abundance of L in virions and its very large size that renders its heterologous expression difficult, and *iii*) the structural flexibility of N and P (see below) [15, 17, 18, 70, 72, 85].

3. Structural Disorder within the Replicative Complex

In the course of the structural and functional characterization of MeV replicative complex proteins, we discovered that the N and P proteins contain long (from 50 to 230 residues) disordered regions possessing sequence features that typify intrinsically disordered proteins [15-18, 70, 72, 85]. Using bioinformatics approaches (see [48] for a review on disorder predictors), we have then further extended these results to *Paramyxovirinae* and, even more generally, to *Mononegavirales* N and P proteins [70]. We found that structural disorder is a widespread property within the replicative complex of these viruses, implying functional relevance. Therefore, *Mononegavirales* constitute an excellent model system for the study of the functional role of disordered regions.

Intrinsically disordered or intrinsically unstructured proteins (IUPs) are functional proteins that fulfill essential biological functions while lacking highly populated constant secondary and tertiary structure under physiological conditions, [41, 44, 51, 123, 124, 127, 131].

The notion that protein function relies on a precise 3D structure constitutes one of the paradigms of biochemistry [3] and is deeply inscribed in our language. During the last two decades, the growing numbers of protein structures determined by X-ray crystallography and by NMR has shifted the attention of structuralists away from the numerous proteins that possess biological functions despite their lack of a precise 3D structure [38, 40-42]. The first call to recognition of proteins containing large disordered regions is very recent [138]. Since then, the number of publications dealing with structural disorder continues to grow. However, the notion of a tight dependence of protein function on a precise 3D structure is still deeply anchored in most structuralists' mind, even though the role of flexible regions within proteins is recognized and despite the familiarity of crystallographers with crystal disorder (both static and dynamic). The reasons for this lack of emphasis upon structural disorder are multiple. Firstly, IUPs have been long unnoticed because researchers encountering examples of structural disorder mainly ascribed it to errors and artifacts and, as such, purged them from papers and reports. Secondly, structural disorder is hard to conceive and classify. Thirdly, IUPs have been neglected because of the perception that a limited amount of mechanistic data can be derived from their study.

Although there are IUPs that carry out their function while remaining disordered all the time (e.g. entropic chains) [41], many of them undergo an unstructured-to-structured transition upon binding to their physiological partner(s), a process termed induced folding [43, 53, 128]. The term induced folding is often associated with a notion of gain of regular secondary structure. However, coupled folding and binding can also occur without such dramatic structural transitions [18], and IUPs tend to conserve their overall extended conformation even after binding to their targets [123]. Thus, induced folding is characterized by a decrease in flexibility of the IUP due to selection by the partner of a particular conformer out of the otherwise numerous conformations that the free IUP can adopt in solution.

The functional importance of disorder is inherent in protein flexibility. In particular, an increased plasticity would *i*) enable binding of numerous structurally distinct targets and *ii*) provide the ability to overcome steric restrictions, enabling larger interaction surfaces in protein-protein and protein-ligand complexes than those obtained with rigid partners, and *iii*) allow protein interactions to occur with both high specificity and low affinity [38, 40-42, 44, 51, 58, 130, 138]. However, although the decrease in conformational entropy that accompanies binding coupled to folding events is thought to lead to a combination of high specificity and weak affinity, this latter point awaits a systematic comparison between experimentally determined K_D values of IUPs and those of structured proteins. Indeed, a considerable number of IUPs display rather high affinities for their targets: this is for example the case of the C-terminal domain of the MeV N [18] (see also below), of SNAP-25 [20], and of the cyclin-dependent kinase inhibitors p21 and p27 [19, 77, 78].

Regions lacking specific 3D structure have been so far associated with approximately 30 distinct functions, including nucleic acid and protein binding, display of phosphorylation sites and of proteolysis sites, prevention of interactions by means of excluded volume effects, and molecular assembly. Most proteins containing disordered regions are involved in signaling and regulation events that generally imply multiple partner interactions (see [26, 38, 39, 67, 123] for reviews). For instance, it has been estimated that approximately 80% of cancer-associated proteins and 60% of cardiovascular-disease associated proteins possess disordered regions [67].

IUPs possess peculiar sequence properties that allow them to be distinguished from globular proteins. In particular, *i*) they are generally enriched in amino acids preferred at the surface of globular proteins (i.e. A, R, G, Q, S, P, E and K) and depleted in W, C, F, I, Y, V, L and N [137], *ii*) they possess a distinct combination of a high content of charged residues and of a low content in hydrophobic residues [128], *iii*) they possess a low predicted secondary structure content and *iv*) they tend to have a low sequence complexity (i.e. they make use of fewer types of amino acids) and *v)* they often have a high sequence variability. This has allowed an estimation of the occurrence of disorder in biological systems. A recent study has shown that between 7 and 35% of proteins from unicellular organisms are predicted to contain long (i.e. >40 residues) disordered regions based upon sequence analysis. This percentage is further increased to 35-60% for multicellular organisms [42]. Successively, it has been estimated that 5% of *E. coli* proteins, 23% of *A. thaliana* proteins and 28% of *M. musculus* proteins are mainly disordered (i.e. they posses more disordered than ordered regions) [99]. More recently, the group of Tompa showed that 50-60% of the *Saccharomyces cerevisiae* proteome contains at least one long (>30) disordered segment [125], and the group of Galzitskaya published a study showing that 12% of eukaryotic proteins would be fully disordered [13]. The increased prevalence of disorder in higher organisms is related to an increased need for cell regulation and signaling. In addition, analysis of the VaZYMolO database, which contains 1683 protein sequences from RNA viruses [50], shows that 6% of these sequences possess at least a disordered region of more than 20 residues in length (Longhi *et al.*, unpublished data). In support of the abundance of disorder within viral proteins, a recent study published by the group of Dunker and Uversky shows that viruses and eukaryota have ten times more conserved disorder (roughly 1%) than archaea and bacteria (0.1%) [25].

4. Structural Organization of the Nucleoprotein

Deletion analyses and electron microscopy studies have shown that *Paramyxoviridae* nucleoproteins are divided into two regions: a structured N-terminal moiety, N_{CORE} (aa 1-400 in MeV), which contains all the regions necessary for self-assembly and RNA-binding [5, 22, 30, 71, 74, 83, 91, 92], and a C-terminal domain, N_{TAIL} (aa 401-525 in MeV), which is intrinsically unstructured [85] (Figure 4A). N_{TAIL} protrudes from the globular body of N_{CORE} and is exposed at the surface of the viral nucleocapsid [62, 63, 71]. N_{TAIL} contains the regions responsible for binding to P in both N°-P and N^{NUC}-P complexes [5, 74, 83, 85].

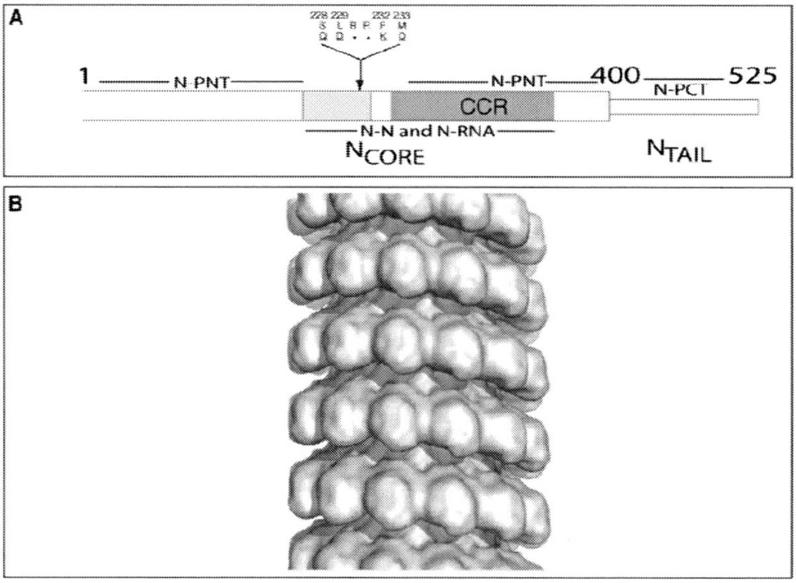

Figure 4. (A) Organization of MeV N. The approximate location of N-N, N-P and RNA-binding sites is indicated. The central conserved region (CCR) (dark grey) and the additional region involved in oligomerization (aa 189-239, light grey) are also shown. The positions targeted for mutagenesis (see text) are shown by an arrow. Wild-type residues are shown in the top position, while mutated residues are shown below. The location of the N_{CORE}-PNT sites (aa 4-188 and 304-373), as reported by [5], is also shown. (B) Reconstruction of MeV nucleocapsid as obtained from cryo-electron microscopy. Data were kindly provided by David Bhella (MRC, Glasgow, UK). The picture was drawn using Pymol (DeLano, W.L. The PyMOL Molecular Graphics System (2002) DeLano Scientific, San Carlos, CA, USA,. http://www.pymol.org).

4.1. The Structured N_{CORE} Domain

N_{CORE} contains all the regions necessary for self-assembly and RNA-binding, since nucleoproteins composed only of the core region can encapsidate neosynthesized RNA into nucleocapsid-like particles. Within N_{CORE}, deletion studies have failed to identify independent, modular domains, but have identified regions involved in the N-N interaction. The region spanning aa 258-357, called the Central Conserved Region (CCR), is well conserved in sequence, and mainly hydrophobic (Figure 4A). Several studies have shown that it constitutes one of the regions involved in self-association and in RNA binding [71, 83]. The location of the RNA-binding site(s) within N_{CORE} is unknown. When denatured, such as in Northwestern blots, N does not bind RNA indicating that the RNA-binding site is probably formed by maturation of N during encapsidation [81]. In agreement, all mutants in which self-association is impaired do not package RNA [71, 92]. In the closely related SeV (a *Respirovirus* member), more subtle mutations in the region 360-375 were found that did neither disrupt RNA-binding nor the morphology of N, but rendered N inactive in replication [92]. The author suggested that particular residues in this region might be involved in binding the leader sequence of SeV RNA, versus non-specific RNA. However, it is more likely that mutations within this region could affect the ability of N to interact with PNT, thereby preventing either formation or proper conformation of the N°-P complex.

Beyond the CCR, another region of N-N interaction was found within residues 189-239 of MeV N, as deletion of this region led to a nucleoprotein variant form that can still bind P but has lost its ability to self-assemble [5]. In further support for a role of this additional region in N assembly, two point substitutions thereof (namely S228Q, L229D) impair self-association of N and RNA binding without affecting neither the overall secondary structure content, nor the gross domain organization and the ability to bind P [71]. Because of its inability to form herringbone-like structures, this N variant is a promising candidate for crystallographic studies.

Indeed, a major hurdle to X-ray crystallography techniques is the strong self-assembly of N to form large nucleocapsids with a broad size distribution when expressed in heterologous systems such as mammalian cells [116], bacteria [136], or insect cells [7]. Because of this property, *Paramyxoviridae* nucleoproteins have resisted so far to high-resolution structure determination. Due to variable helical parameters, the recombinant or viral nucleocapsids are also difficult to analyze using electron microscopy coupled to image analysis. Despite these technical drawbacks, elegant electron microscopy studies by two independent groups led to real-space helical reconstruction of MeV nucleocapsids [8, 112] (Figure 4B). These studies pointed out a considerable conformational flexibility and also showed that removal of the disordered N_{TAIL} domain leads to increased nucleocapsid rigidity, with significant changes in both pitch and twist (see [85, 112] and Chapter 2).

Conversely, high-resolution structural data are available for two *Rhabdoviridae* members, namely the rabies virus and the VSV [1, 57]. The nucleoprotein of these viruses consists of two lobes and possesses an extended terminal arm that makes contacts with a neighboring N monomer. The RNA is tightly packed in a cavity between the two N lobes. N establish contacts with the sugar and phosphate moiety of nucleotides via basic residues, in agreement with previous studies showing that the phosphate moieties of encapsidated RNA are not accessible to the solvent [68]. In both nucleoproteins the RNA is not accessible to the solvent. Thus, it has to partially dissociate from N to become accessible to the polymerase.

Functional and structural similarities between the nucleoproteins of *Rhabdoviridae* and *Paramyxoviridae* are well established. In particular, they share the same organization in two well-defined regions, N_{CORE} and N_{TAIL}, and in both families the CCR is involved in RNA-binding and self-assembly of N [76, 92]. Moreover, incubation of the rabies virus nucleocapsid with trypsin results in the removal of the C-terminal region (aa 377-450) [76].

This N_{TAIL}-free nucleocapsid is no longer able to bind to P, thus suggesting that in *Rhabdoviridae*, N_{TAIL} plays a role in the recruitment of P as in the case of *Paramyxoviridae* [111]. However, contrary to MeV, the rabies virus N_{TAIL} domain is structured [1]. Presently, it is not known whether *Rhabdoviridae* and *Paramyxoviridae* nucleoproteins share the same bilobal morphology. In the case of MeV N, bioinformatics analyses suggest that N_{CORE} is organised into two subdomains (aa 1-130 and aa 145-400) (see [14, 48]) separated by a hypervariable, antigenic loop (aa 131-149) [55] that probably fold cooperatively. It is tempting to speculate that this organization could indicate that MeV N_{CORE} could have a bilobal morphology.

Several features distinguish the N proteins of *Rubulaviruses* (a genus within the *Paramyxoviridae* family) from those of *Respiroviruses* and *Morbilliviruses*. While in *Respiroviruses* and *Morbilliviruses* N forms large cytoplasmic inclusion bodies when expressed alone in mammalian cells, *Rubulavirus* N does not, unless it is co-expressed with P [102]. Nevertheless, simian virus 5 (a *Rubulavirus* member) N expressed in insect cells binds to cellular RNA and forms nucleocapsid-like particles even in the absence of P [7]. Finally, the *Rubulavirus* N^{NUC}-P binding site is located within N_{CORE}, with no binding to N_{TAIL} being observed [74, 93].

Although within the MeV N^{NUC}-P complex, the N region responsible for binding to P is located within N_{TAIL}, the N°-P complex involves an additional interaction between N_{CORE} and the disordered N-terminal domain of P (PNT) (Figure 3). Within the N°-P complex, P to N binding is mediated by the dual PNT-N_{CORE} and PCT-N_{TAIL} interaction [27] (Figure 3). Studies on SeV suggested that an N°-P complex is absolutely necessary for the polymerase to initiate encapsidation but that in the presence of large amounts of soluble N alone, the polymerase can carry out replication of preinitiated chains [4]. Therefore, formation of the N°-P complex would have at least two separate functions: *i)* prevent illegitimate self-assembly of N, and *ii)* allow the polymerase to deliver N to the nascent RNA to initiate replication.

The regions within N_{CORE} responsible for binding to PNT within the N°-P complex have been mapped to residues 4-188 and 304-373, with the latter region being not strictly required for binding and rather favoring it [5] (Figure 4A). However, precise mapping of such regions is hard because N_{CORE} does not have a modular structure, and consequently it is difficult to distinguish between gross structural defects and specific effects of deletions.

4.2. The Intrinsically Unstructured N_{TAIL} Domain

In *Morbilliviruses,* N_{TAIL} is responsible for binding to P in both N°-P and N^{NUC}-P complexes [5, 74, 83, 85]. Within the N^{NUC}-P complex, N_{TAIL} is also responsible for the interaction with the polymerase (L-P) complex [5, 74, 83, 85]. Contrary to N_{CORE}, the amino acid sequence of N_{TAIL} is hyper-variable within *Morbillivirus* members. In addition, N_{TAIL} is hyper-sensitive to proteolysis [71] and is not readily visualized in electron microscopy. As these latter features are hallmarks of intrinsic disorder, we analyzed the sequence properties of N_{TAIL} in order to assess whether they conform to those of IUPs [41]. The amino acid composition of N_{CORE} does not deviate from the average composition of proteins found in the

Protein Data Bank (PDB). Conversely, the N_{TAIL} region has a peculiar sequence composition. It is depleted in "order promoting" residues (W, C, F, Y, I, L) and enriched in "disorder promoting" residues (R, Q, S et E) [85]. Moreover, N_{TAIL} is predicted to be intrinsically disordered by both PONDR [107], the method based on the mean hydrophobicity/mean net charge ratio [129] (see [15, 70]), as well as by other disorder predictors such as IUPred [36, 37] (Longhi *et al.*, unpublished data).

In order to experimentally assess the disordered nature of N_{TAIL} we expressed this protein domain in *E. coli*. We then investigated the structural properties of N_{TAIL} by using a combination of several biochemical and biophysical approaches (see [104] for a review on methods to assess structural disorder and induced folding). After purification from the soluble fraction of *E. coli*, N_{TAIL} displays an abnormally slow migration in SDS-PAGE, with an apparent molecular mass of 20 kDa (expected mass: 15 kDa) [85]. This anomalous behavior is related to a rather high content in acidic residues, a characteristic that has been already documented for other IUPs [122], such as, for instance the MeV PNT [72], as well as for all *Paramyxovirinae* P proteins [81].

Analysis of the hydrodynamic properties of N_{TAIL} shows that this domain has a very extended shape in solution, with a hydrodynamic radius (Stokes radius) of 27 Å [85]. The theoretical expected Stokes radii for a globular and an unfolded protein of the same molecular mass as that of N_{TAIL} are 19 and 35 Å, respectively. Thus, the experimentally observed Stokes radius is not compatible with the value expected for a globular protein, but rather with that expected either for an unfolded protein or for a structured dimer. 2D-NMR (bi-dimensional Nuclear Magnetic Resonance) and circular dichroism (CD) experiments allowed us to discriminate between these two hypotheses, since both 2D-NMR (Figure 5A) and CD (Figure 5B) spectra of N_{TAIL} are typical of an unfolded protein [85].

The absence of a globular core has been further demonstrated by limited proteolysis experiments, which showed that N_{TAIL} is fully exposed to the solvent (Longhi *et al.*, unpublished data).

N_{TAIL} has also been studied by Small Angle X-ray Scattering (SAXS). This technique is particularly well adapted to study flexible, low compactness or even extended macromolecules in solution. It provides low-resolution structural data, and gives access to the mean particle size (radius of gyration, R_g) as well as to the maximal intramolecular distance (D_{MAX}). These two parameters give information on the degree of compactness of the molecule, and the latter gives an idea of the maximal degree of extension reached by the molecule in solution. In the case of N_{TAIL}, the obtained values indicate that this protein domain is not globular, yet conserves a certain extent of compactness [85]. Likewise, the spectroscopic properties of N_{TAIL} as observed by CD studies, indicate that it possesses some residual secondary structure [85]. Thus N_{TAIL}, while being an IUP, conserves some residual structure that characterizes the "premolten globule" subfamily [127]. Premolten globules are typified by a conformational state intermediate between a random coil and a molten globule, where premolten globules are more compact (but still less compact than globular or molten globule proteins) [41, 127]. In solution they possess a certain degree of residual compactness due to the presence of residual and fluctuating secondary and tertiary structures.

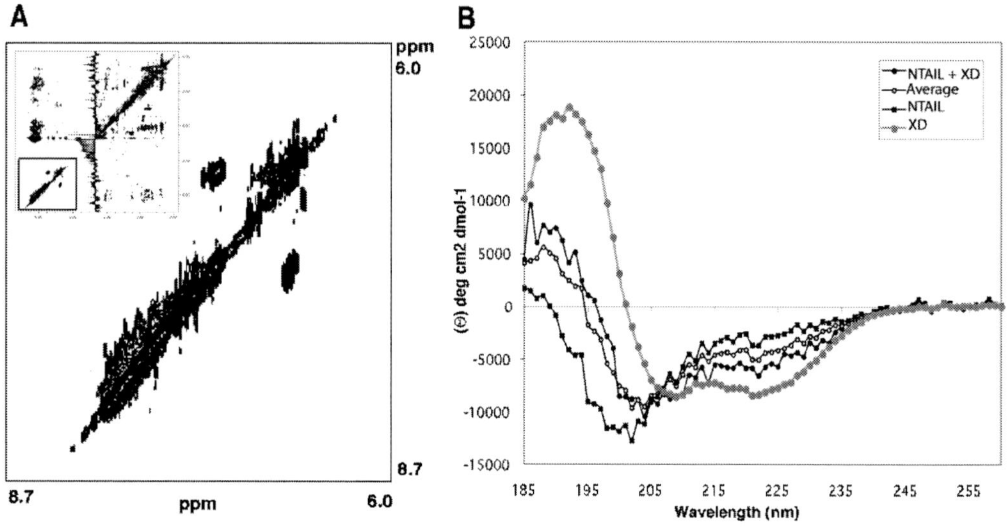

Figure 5. (A). 2D-NMR spectrum of N_{TAIL} between 7.8 and 8.7 ppm. The very low spread of the amide protons resonance frequencies is typical of proteins with no stable secondary structure. The inset shows the 2D-NMR spectrum of N_{TAIL} between 0 and 8.7 ppm. (B). CD spectra of N_{TAIL}, of XD and of a mixture of N_{TAIL} and XD at a 1:2 molar ratio. The CD spectrum of N_{TAIL} is typical of an unfolded protein, as seen by the nil ellipticity value at 185 and by the largely negative ellipticity value at 198 nm. The CD spectrum of XD is typical of a folded protein with a predominant α-helical content. The theoretical average spectrum, computed by averaging the individual spectra, is also shown. It corresponds to the theoretical CD spectrum expected in case no structural transition takes place. The deviation of the experimentally observed CD spectrum obtained with the N_{TAIL}-XD mixture from the theoretical average spectrum indicates a coil to α-helix transition (see appearance of minima at 208 and 222 nm, and increase in the ellipticity in the 185-195 nm region). Data of panel A were taken from [85], while those of panel B were taken from [69].

As for the functional implications, It has been proposed that the residual intramolecular interactions that typify the premolten globule state may enable a more efficient start of the folding process induced by a partner [53, 78, 122].

CD studies in the presence of increasing concentrations of tri-fluoro-ethanol (TFE) pointed out a clear gain of α-helicicity [85]. TFE is an organic solvent that mimics the hydrophobic environment experienced by proteins in protein-protein interactions. It is widely used as a probe for regions that have a propensity to undergo induced folding. In agreement with these results, CD studies indicated that N_{TAIL} undergoes an α-helical transition in the presence of PCT, whereas no structural transition is observed in the presence of PNT [85]. Using computational approaches, an α-helical *M*olecular *R*ecognition *E*lement (α-MoRE, aa 488-499 of N) has been identified within N_{TAIL}. MoREs are regions within IUPs that have a certain propensity to bind to a partner and thereby to undergo induced folding [54, 100]. In the case of N_{TAIL}, an α-helical transition is predicted [15], in agreement with the results of secondary structure predictions that identified an α-helix within residues 488-504 as the sole secondary structure element [85]. The role of the α-MoRE in binding to P and in induced folding has further been confirmed by spectroscopic and biochemical experiments carried out on a truncated N_{TAIL} form devoid of the 489-525 region [15]. The P region responsible for the interaction with N_{TAIL} and the induced folding of this latter has been mapped to the C-terminal module (XD, aa 459-507) of P [69] (Figure 5B). The crystal structure of XD has been solved and consists of a triple α-helical bundle. This structure, beyond representing the first solved structure of a MeV protein, sheds light on the molecular mechanism of the induced folding of N_{TAIL}. The surface of XD delimited by helices α2 and α3 contains a large hydrophobic cleft that could accommodate the α-MoRE of N_{TAIL}, promoting the induced folding of the latter (Figure 6).

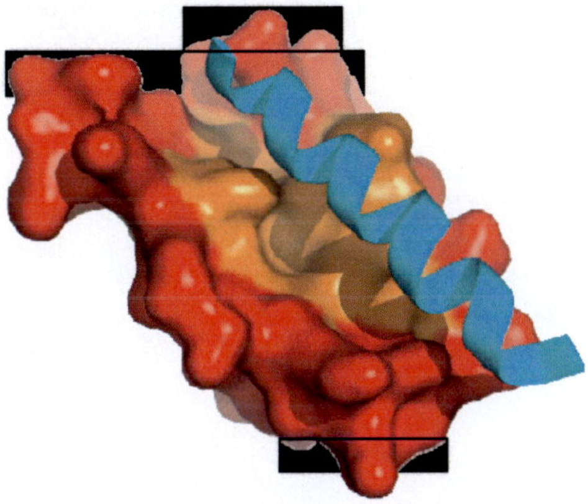

Figure 6. Model of the complex between XD and the α-MoRE of N_{TAIL} according to [69]. The picture was drawn using Pymol (DeLano, W.L. The PyMOL Molecular Graphics System (2002) DeLano Scientific, San Carlos, CA, USA,. http://www.pymol.org).

By combining structural and biochemical data, we built a model of the interaction between XD and the α-MoRE of N_{TAIL}. According to this model, burying of hydrophobic residues would provide the driving force to induce folding of the α-MoRE (Figure 6), thus leading to a pseudo-four helix arrangement occurring frequently in nature [69]. This model has been recently validated by Kingston and co-workers who solved the crystal structure of a chimeric form mimicking this complex [73].

SAXS studies allowed us to derive a low-resolution model of the N_{TAIL}-XD complex. This model (Figure 7) shows that most of N_{TAIL} (aa 401-488) remains disordered within the complex. The absence of a protruding shape from the bulky region (Figure 7) indicates that the C-terminus of N_{TAIL} does not point towards the solvent and must rather be part of the globular region, thus suggesting that, beyond the α-MoRE, the C-terminus may be also involved in binding to XD [18]. Indeed, spectroscopic (CD, fluorescence) studies carried out on truncated forms of N_{TAIL} devoid of regions conserved within *Morbillivirus* N proteins (namely, Box1-Box3), pointed out the involvement of the C-terminus of N_{TAIL} in the interaction with XD [18]. Surface plasmon resonance (BIAcore) studies carried out with truncated N_{TAIL} proteins further supported a role for Box3 (aa 517-525) in binding, where removal of either Box3 alone or Box2 (aa 488-506) *plus* Box3 results in a strong increase (three orders of magnitude, K_D 10 μM) in the equilibrium dissociation constant [18]. When synthetic peptides mimicking Box1 (aa 401-420), Box2 and Box3 were used, we found that Box2 peptide displays an affinity towards XD (K_D of 20 nM) similar to that of N_{TAIL} (K_D of 80 nM) consistent with the role of Box2 as the primary binding site (Longhi and Oglesbee, unpublished data). Interestingly however, Box3 peptide exibits a significantly lower affinity for XD (K_D of approximately 1 mM) as compared to that of Box2 peptide (Longhi and Oglesbee, unpublished data).

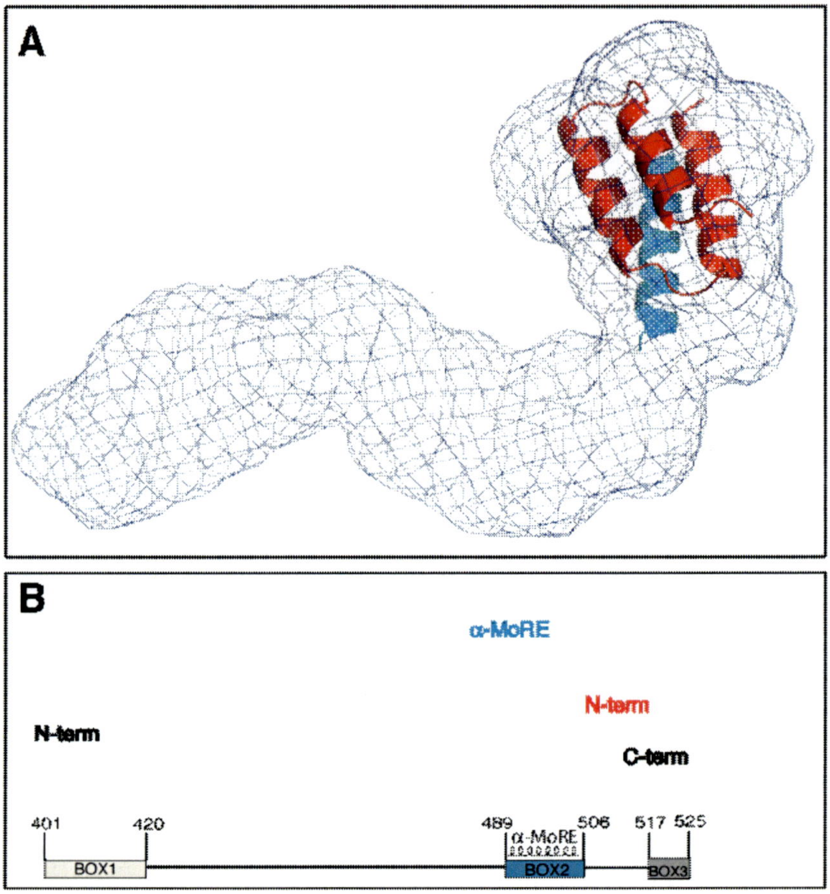

Figure 7. (A) Global shape of the N_{TAIL}-XD complex. The crystal structure of the chimera between XD (red) and the α-MoRE (blue) is shown. The picture was drawn using Pymol (DeLano, W.L. The PyMOL Molecular Graphics System (2002) DeLano Scientific, San Carlos, CA, USA,. http://www.pymol.org). (B) Low resolution model of the N_{TAIL}-XD complex showing that the 401-488 region of N_{TAIL} is disordered and exposed to the solvent, while the α-MoRE and the C-terminus are packed against XD.

The discrepancy between the data obtained with N_{TAIL} truncated proteins and with peptides can be accounted for by assuming that Box3 would act only in the context of N_{TAIL} and not in isolation. Thus, according to this model, Box3 and Box2 would be functionally coupled in the binding of N_{TAIL} to XD. We can speculate that burying the hydrophobic side of the α-MoRE in the hydrophobic cleft formed by helices α2 and α3 of XD would provide the primary driving force in the N_{TAIL}-XD interaction and that Box3 would act as a "clamp". The combined interaction of Box2 and Box3 with XD would provide further stabilization of the complex as compared to a complex devoid of Box3. We tentatively propose that binding to XD might take place through a sequential mechanism which could involve binding and folding of Box2, followed by binding of Box3 (see Figure 8).

Heteronuclear NMR (HN-NMR) experiments allowed us to precisely estimate of the number of residues involved in the interaction with XD, as well as to elucidate the nature of the structural transition [18].

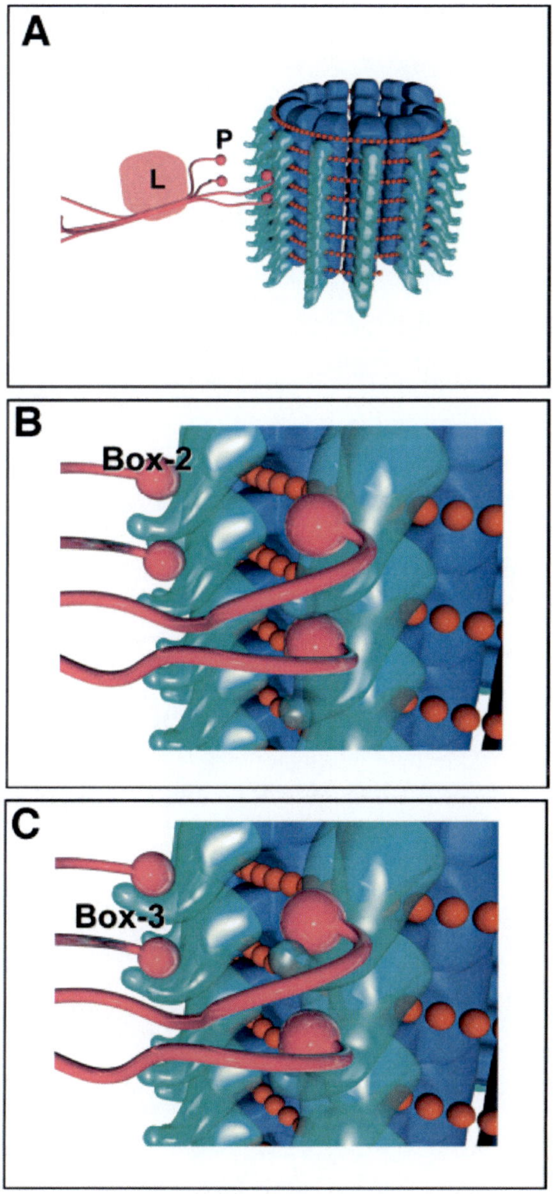

Figure 8. (A). Schematic representation of the P-L complex (magenta) bound to the nucleocapsid. The N_{CORE} region is represented in blue, while the disordered N_{TAIL} domain protruding form the surface of the nucleocapsid is shown in light green. The viral RNA is represented in red, with each sphere corresponding to a nucleotide. The RNA has been represented embedded in the middle of N and as located at the exterior of the nucleocapsid by analogy with *Rhabdoviridae* N-RNA complexes [1, 57]. (B) and (C). Schematic representation of the sequential mechanism by which XD interacts with N_{TAIL} showing binding of Box2 (B) followed by binding of Box3 (C). Courtesy of Tim Vojt, College of Veterinary Medicine, The Ohio State University, USA.

Only 18 N_{TAIL} residues (out of 125) participate in the interaction with XD. Among them, 11 undergo a coil to α-helix transition, and very likely belong to the 486-503 region (encompassing the α-MoRE), in agreement with the NMR studies performed by Kingston *et al.* [73]. The seven other residues undergo a less dramatic displacement that reflects only a change in their chemical environment with no gain of regular secondary structure [18]. HN-NMR experiments carried out with a truncated form of N_{TAIL} definitely showed that these latter residues map within the C-terminus (aa 517-525) of N_{TAIL} [18]. Altogether, these studies showed that binding of N_{TAIL} to XD involves different types of interactions. While the α-MoRE gains α-helical structure upon binding to XD, the C-terminus de N_{TAIL} does not gain any regular secondary structure elements. Nevertheless, it plays a crucial role in stabilizing the complex [18].

We also characterized the structural disorder and the induced folding of N_{TAIL} by using site-directed spin-labeling electron paramagnetic resonance (EPR) spectroscopy. EPR spectroscopy specifically detects unpaired electrons. EPR sensitive reporter groups (spin labels or spin probes) can be introduced into biological systems *via* site-directed spin-labeling (SDSL). The basic strategy of SDSL involves the introduction of a paramagnetic nitroxide side chain at a selected protein site. This is usually accomplished by cysteine-substitution mutagenesis, followed by covalent modification of the unique sulphydryl group with a selective nitroxide reagent, such as the methanethiosulphonate (MTSL) derivative (see [9, 47, 65] for reviews). The spin label is then observed by EPR spectroscopy. The mobility of the spin label reflects the local mobility of residues in the proximity of the radical. The ratio of the peak-to-peak amplitudes of the low field and central field lines (named h(+1)/h(0) herein after) is an indicator of the spin label mobility (Figure 9A). In particular, this ratio decreases with decreasing mobility (Figure 9A). Variations in the radical mobility can therefore be monitored in the presence of partners, ligands, or organic solvents.

We thus designed, purified and labeled with a nitroxide paramagnetic probe four single-site N_{TAIL} mutants (S407C, S488C, L496C and V517C) (Figure 9B). By comparing the radical mobility under native and denaturing (8M urea) conditions, we unveiled the existence of some residual secondary and/or tertiary structure in the proximity of positions 488 and 496 [89]. This may reflect the predominance of an α-helical conformation among the highly fluctuating conformations sampled by unbound N_{TAIL}, in agreement with previous biochemical and spectroscopic data that mapped the region involved in the α-helical induced folding to residues 489-506 [15]. That the conformational space of MoREs [100] in the unbound state is restricted by their inherent conformational propensities, thereby reducing the entropic cost of binding, has already been proposed [53, 78, 113, 122].

We then monitored the gain of rigidity that N_{TAIL} undergoes in the presence of either TFE or XD. Using TFE, we showed that the C-terminal region of N_{TAIL} "resists" to the gain of structure even at TFE concentrations as high as 40% [89]. This latter observation points out the relevance of studies making use of TFE to infer information about protein structural propensities.

The mobility of the spin label grafted at positions 488, 496 and 517 is significantly reduced upon addition of XD, contrary to that of the spin label bound to position 407, which was unaffected [89]. The drop in the mobility of the spin labels grafted at positions 488 and 496 triggered by XD is more pronounced than that induced by TFE [89]. This difference can likely be ascribed either to the ability of XD to stabilize the formation of the α-MoRE (with this latter being only transiently populated by the unbound form of N_{TAIL} in the presence of TFE), or to a restrained motion of the backbone due to the presence of XD.

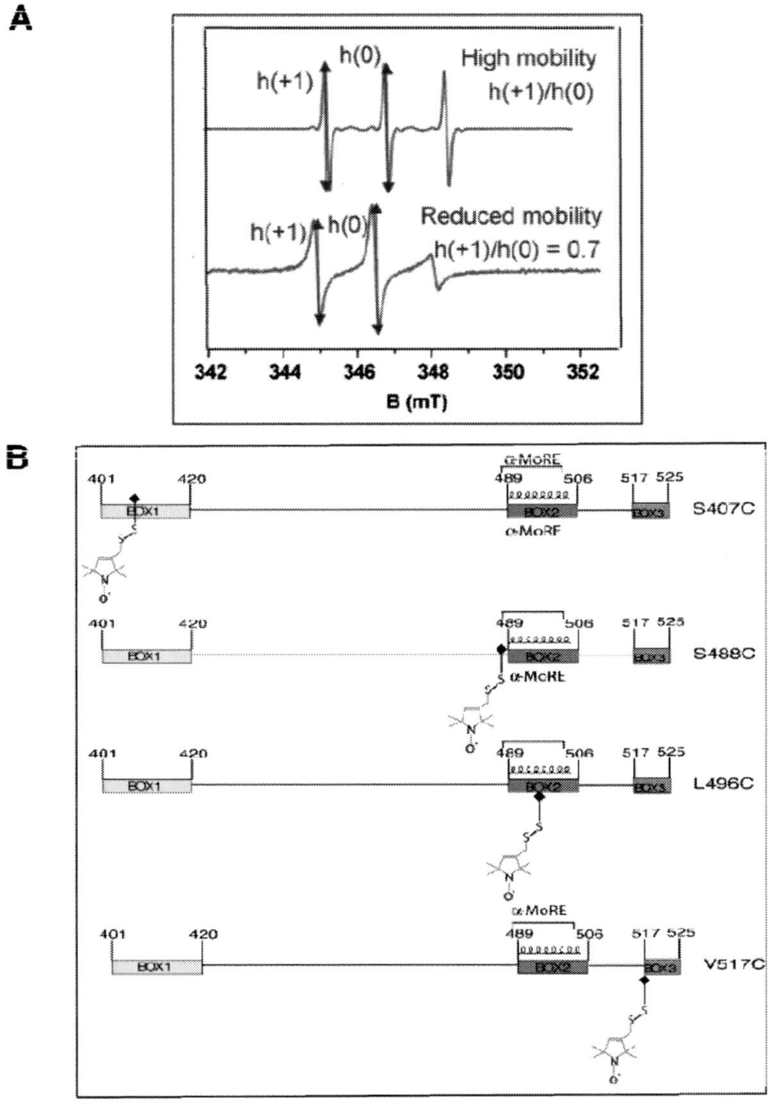

Figure 9. (A) EPR spectra of spin-labeled S488C N_{TAIL} protein in the absence (top) or presence of 10% TFE. (B) Schematic representation of the four spin-labeled N_{TAIL} proteins. The nitroxide spin label and the α-MoRE are shown. Data of panel A were taken from [89].

Furthermore, the EPR spectra of spin-labeled S488C and L496C bound to XD in the presence of 30% sucrose (i.e. under conditions in which the intrinsic motion of the protein becomes negligible with respect to the intrinsic motion of the spin label), are indicative of the formation of an α-helix in the proximity of the spin labels [89]. Furthermore, in the presence of sucrose, a slightly higher mobility is observed for the spin label at position 488 as compared to position 496. This difference may reflect a more constrained motion of the spin label at position 496 as compared to position 488, because of their different location relative to the α-MoRE. That the extremities of α-helices are more mobile than their central parts has already been well documented by SDSL EPR spectroscopy [88].

On the other hand, the addition of XD triggers only a modest reduction in the mobility of the spin label grafted at position 517 with respect to the shifts observed for both spin-labeled S488C and L496C proteins. This observation is consistent with previous data indicating that Box3 does not undergo α-helical folding upon binding to XD [18]. In further support of the lack of α-helical folding of Box3, the EPR spectrum of the V517C-XD complex does not exhibit the typical signature that is observed in the spectra of the S488C-XD and L496C-XD complexes and that is attributable to an α-helix [89].

Finally, by using equilibrium displacement experiments we showed not only the reversibility of the N_{TAIL}-XD interaction, but also the reversibility of the induced folding of the α-MoRE upon dissociation of XD [89]. Indeed, upon dissociation of XD, the spin-labeled S488C and L496C proteins exhibit a spectral signature corresponding to the free form, consistent with a loss of α-helicity [89]. These results represent the first experimental evidence indicating that N_{TAIL} adopts its original pre-molten globule conformation after dissociation of its partner.

This latter point is particularly relevant taking into consideration that the contact between XD and N_{TAIL} within the replicative complex has to be dynamically made and broken to allow the polymerase to progress along the nucleocapsid template during both transcription and replication. Hence, the complex cannot be excessively stable for this transition to occur efficiently at a high rate (see below).

In conclusion using a panel of various physico-chemical approaches, we showed that the interaction between N_{TAIL} and XD implies the stabilization of the helical conformation of the α-MoRE, which is otherwise only transiently populated in the unbound form, and a dramatic reduction in the mobility of Box3 due to selection of a conformer by the partner. Preliminary mapping of Box3 binding sites within XD using HN-NMR, suggests that Box3 may rather establish direct contacts with Box2 (Darbon, Bernard and Longhi, unpublished data). According to this model, binding to XD would cause α-helical folding of the α-More, its embedding in the hydrophobic cleft of XD, followed by a conformational change of Box3, which would move close to Box2. As a result, the chemical environment of Box3 is modified and the overall mobility of the C-terminus is reduced. Experiments aimed at precisely determining the location within the N_{TAIL}-XD complex of Box3 with respect to Box2 are in progress.

5. Functional Role of Structural Disorder of N_{TAIL} for Transcription and Replication

The K_D value between N_{TAIL} and XD is in the 100 nM range [18]. This affinity is considerably higher than that reported by Kingston and co-workers (K_D of 13 μM) and derived from isothermal titration calorimetry studies [74]. A weak binding affinity, associated with a fast association rate, would ideally fulfill the requirements of a polymerase complex which has to cartwheel on the nucleocapsid template during both transcription and replication. However, a K_D in the μM range would not seem to be physiologically relevant considering the low intracellular concentrations of P in the early phases of infection, and the relatively long half life of active P-L transcriptase complex tethered on the NC template, which has been determined to be well over 6 hours [101]. Moreover, such a weak affinity is not consistent with the ability to readily purify nucleocapsid-P complexes using rather stringent techniques such as CsCl isopycnic density centrifugation [96, 105, 106, 117]. A more stable XD-N_{TAIL} complex would be predicted to hinder the processive movement of P along the nucleocapsid template. In agreement with this model, the elongation rate of MeV polymerase was found to be rather slow (three nucleotides/s) [101]. In addition, the C-terminus of N_{TAIL} has been shown to have an inhibitory role upon transcription and replication, as indicated by minireplicon experiments, where deletion of the C-terminus of N enhanced basal reporter gene expression [142]. Deletion of the C-terminus of N also reduces the affinity of XD for N_{TAIL}, providing further support for modulation of XD/N_{TAIL} binding affinity as a basis for polymerase processivity. Thus, Box3 would dynamically control the strength of the N_{TAIL}-XD interaction, by stabilizing the Box2-XD interaction. Removal of Box3 or interaction of Box3 with other partners (see below and Chapter 3), would reduce the affinity of N_{TAIL} for XD thus stimulating transcription. Modulation of XD/N_{TAIL} binding affinity could be dictated by interactions between N_{TAIL} and cellular and/or viral co-factors. Indeed, the requirement for cellular or viral co-factors in both transcription and replication has been already documented in the case of MeV [134], and of other *Mononegavirales* members [46, 60]. Furthermore, in both CDV and MeV the viral transcription and replication are enhanced by the major heat shock protein (hsp72), and this stimulation relies on interaction with N_{TAIL} [141, 142]. These co-factors may serve as processivity or transcription elongation factors and could act by modulating the strength of the interaction between the polymerase complex and the nucleocapsid template (see below and Chapter 3).

N_{TAIL} also influences the physical properties of the nucleocapsid helix that is formed by N_{CORE} [85, 112]. Electron microscopic analysis of nucleocapsids formed by either N or N_{CORE} indicates that the presence of N_{TAIL} is associated with a greater degree of fragility, evidenced by the tendency of helices to break into individual ring structures (Figure 10A). This fragility is associated with evidence of increased nucleocapsid flexibility, with helices formed by N_{CORE} alone forming rods (Figure 10A) (see also [112] and Chapter 2). It is therefore conceivable that the induced folding of N_{TAIL} resulting from the interaction with P (and/or other physiological partners) could also affect the nucleocapsid conformation in such a way as to affect the structure of the replication promoter (Figure 10B). Indeed, the replication promoter, located at the 3' end of the viral genome, is composed of two discontinuous elements building up a functional unit because of their juxtaposition on two successive helical turns [119] (Figure 10B). The switch between transcription and replication could be dictated by variations in the helical conformation of the nucleocapsid, which would result in a modification in the number of N monomers (and thus of nucleotides) *per* turn, thereby disrupting the replication promoter in favor of the transcription promoter (or *vice versa*). Morphological analyses, showing the occurrence of a large conformational flexibility within *Paramyxoviridae* nucleocapsids [7, 8, 95, 96], tend to corroborate this hypothesis (see also Chapters 2 and 3).

Finally, preliminary data indicate that incubation of MeV NCs in the presence of XD triggers unwinding of the NC, thus possibly enhancing the accessibility of genomic RNA to the polymerase complex (Bhella and Longhi, unpublished data). Hence, we propose that the XD-induced α-helical folding of N_{TAIL} could trigger the opening of the two lobes of N_{CORE} thus rendering the genomic RNA accessible to the solvent.

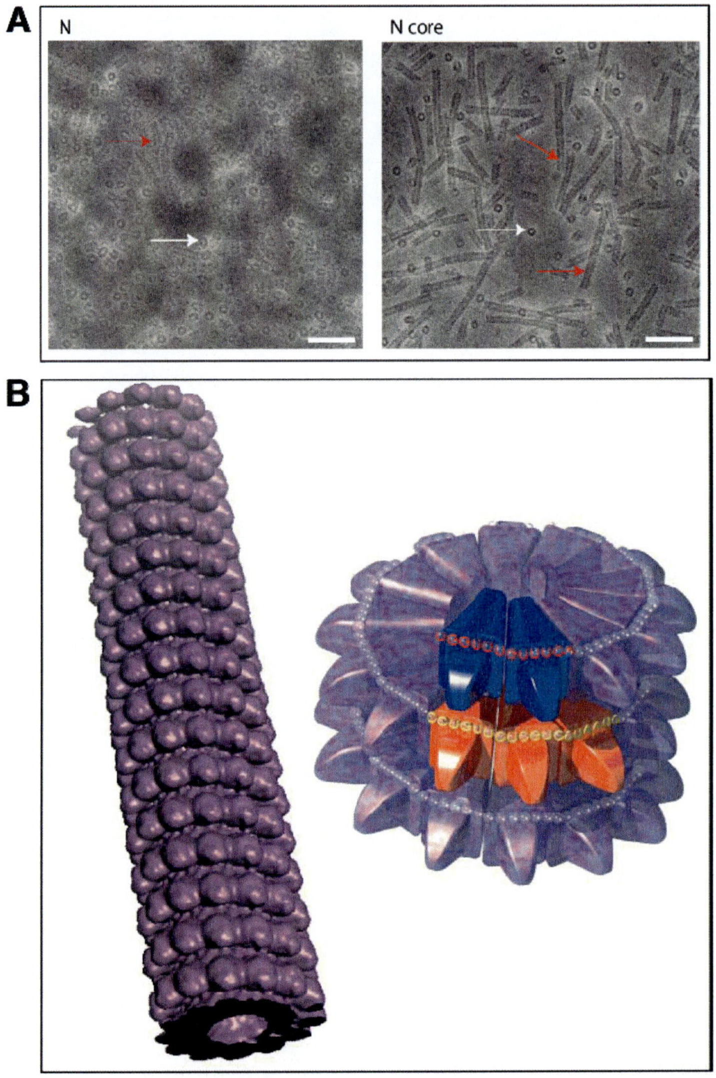

Figure 10. (A) Negative stain electron micrographs of N and N_{CORE}. The bar corresponds to 100 nm. Rings and herringbone structures are indicated by white and red arrows, respectively. (B) Cryo-electron microscopy reconstructions of MeV nucleocapsid (left) and schematic representation of the nucleocapsid (right) highlighting the structure of the replication promoter composed of two discontinuous units juxtaposed on successive helical turns (see regions wrapped by the red and blue N monomers) (courtesy of D. Bhella, MRC, Glasgow, UK). Data of panel A were taken from [85].

Unstructured regions are considerably more extended in solution than globular ones. For instance, MeV PNT has a Stokes radius of 4 nm [72]. However, the Stokes radius only reflects a mean dimension. Indeed,the maximal extension of PNT, as measured by SAXS (Longhi *et al.*, unpublished data) is considerably larger (> 40 nm). In comparison, one turn of the nucleocapsid is 18 nm in diameter and 6 nm high [7]. Thus PNT could easily stretch over several turns of the nucleocapsid, and since P is multimeric, N°-P might have a considerable extension (Figure 11).

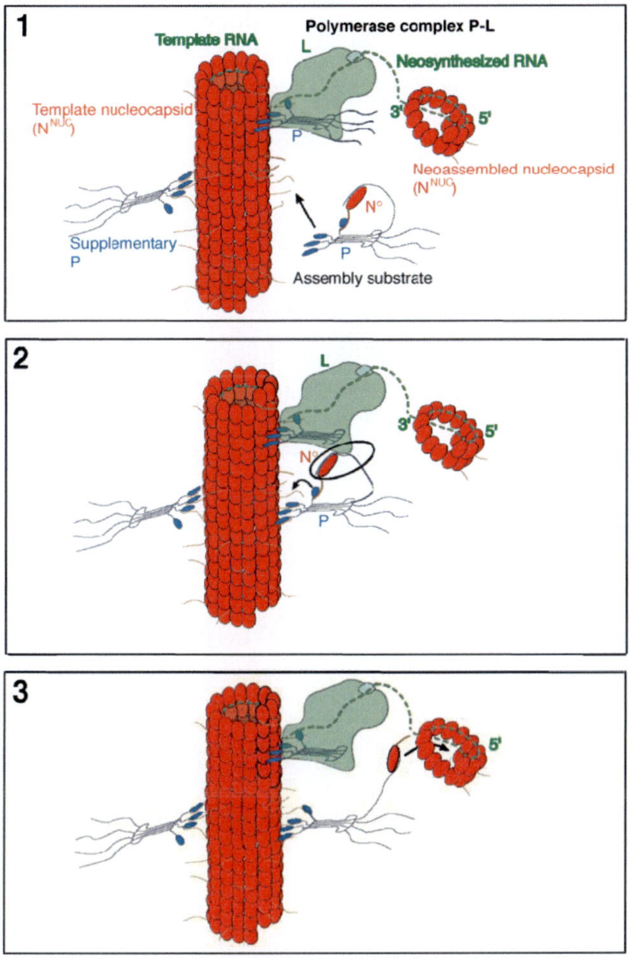

Figure 11. (A) Model of the polymerase complex actively replicating genomic RNA. Disordered regions are represented by lines. The location of the viral RNA (dotted line) within the N^{NUC}-P complex is schematically represented at the interior of the nucleocapsid only for clarity. Within the N^{NUC}-P complex, P is represented as bound to N^{NUC} through three of its four terminal XD arms according to the model of [31]. Only a few N_{TAIL} regions are drawn. PNT regions within the L-P complex have not been represented in panels 2 and 3. The P molecule delivering N° has been represented as distinct from that within the L-P complex in agreement with the results of [126]. The numbering of the different panels indicates the chronology of events. (1) L is bound to a P tetramer. A supplementary P molecule, not bound to L, is also shown (right). The newly-synthesized RNA is shown as already partially encapsidated. (2) The encapsidation complex, N°-P, binds to the nucleocapsid template through three of its four XD arms. The extended conformation of N_{TAIL} and PNT would allow the formation of a tripartite complex between N°, P, and the polymerase (circled). It is tempting to imagine that the proximity of the polymerase (or an unknown signal from this latter) may promote the release of N° by XD, thus leading to N° incorporation within the assembling nucleocapsid. The N° release would also lead to cartwheeling of the L-P complex through binding of the free XD arm onto the nucleocapsid template (see arrow) as in the model of [31]. (3) PNT delivers N° to the newly-assembled nucleocapsid (see arrow).

In the same vein, it is striking that SeV and MeV PCT, which interacts with the intrinsically disordered N_{TAIL} domain, comprises a flexible linker [6, 11, 85, 86]. This certainly suggests the need for a great structural flexibility. This flexibility could be necessary for the tetrameric P to bind several turns of the helical nucleocapsid. Indeed, the promoter signals for the polymerase are located on the first and the second turn of the SeV nucleocapsid [119].

Likewise, the maximal extension of N_{TAIL} in solution is of 13 nm [85]. The very long reach of disordered regions could enable them to act as linkers and to tether partners on large macromolecular assemblies. Accordingly, one role of the tentacular N_{TAIL} projections in actively replicating nucleocapsids could be to put into contact several proteins within the replicative complex, such as the N°-P and the P-L complexes (see Figure 11).

In *Respiroviruses* and *Morbilliviruses*, PNT, which is also disordered [70], contains binding sites for N° [33, 61, 118] and for L [32, 34, 118]. This pattern of interactions among N°, P and L, mediated by unstructured regions of either P or N, suggests that N°, P, and L might interact simultaneously at some point during replication. Notably, the existence of a N-P-L tripartite complex has been recently proved by co-immunoprecipitation studies in the case of VSV, where this tripartite complex constitutes the replicase complex, as opposed to the L-P binary transcriptase complex [59].

We propose that during replication, the extended conformation of PNT and N_{TAIL} is key to allowing contact between the assembly substrate (N°-P) and the polymerase complex (L-P), forming a tripartite N°-P-L complex (Figure 11). This model emphasizes the plasticity of intrinsically disordered regions, which might give a considerable reach to the elements of the replicative machinery.

Interestingly, there is a striking parallel between the N_{TAIL}-XD interaction and the PNT-N_{CORE} interaction (Figure 11). Both interactions are not stable by themselves and must be strengthened by the combination of other interactions. This might ensure easy breaking and reforming of interactions. The relative weak affinity that typifies interacting disordered regions, together with their ability to establish contacts with other partners serving as potential regulators, would ensure dynamic breaking and reforming of interactions. This would result in transient, easily modulated interactions. One can speculate that the gain of structure of N_{TAIL} upon binding to XD could result in stabilization of the N-P complex. At the same time, folding of N_{TAIL} would result in a modification in the pattern of solvent-accessible regions resulting in the shielding of specific regions of interaction. As a result, N_{TAIL} would no longer be available for binding to its other partners. Although induced folding likely enhances the affinity between interacting proteins, the dynamic nature of these interactions could rely on *i)* the intervention of viral and/or cellular co-factors modulating the strength of such interactions, and *ii)* the ability of the IUP to establish weak affinity interactions through residual disordered regions.

Finally, as binding of N_{TAIL} to XD allows tethering of the L protein on the NC template, the N_{TAIL}-XD interaction is crucial for both viral transcription and replication. Moreover, as neither N_{TAIL} not XD have cellular homologues, this interaction is an ideal target for antiviral inhibitors. *In silico* screening of small compounds for their ability to bind to the hydrophobic cleft of XD is in progress. A few candidate molecules are being tested for their ability to bind to XD and to prevent interaction with N_{TAIL} using HN-NMR (Morelli, Guerlesquin and Longhi, unpublished data).

6. N_{TAIL} and Molecular Partnership

The disordered nature of N_{TAIL} confers to this N domain the ability to adapt to various partners and to form complexes that are critical for both transcription and replication. Indeed, thanks to its exposure at the surface of the viral nucleocapsid, N_{TAIL} establishes numerous interactions with various viral partners, including P, the P-L complex and possibly the matrix protein [29]. Beyond viral partners, N_{TAIL} also interacts with several cellular proteins, including the interferon regulatory factor 3 (IRF-3) [121], the heat-shock protein hsp72 [141, 142], the cell protein responsible for the nuclear export of N [110], and possibly components of the cell cytoskeleton [35, 90]. Moreover, N_{TAIL} within viral nucleocapsids released from infected cells also binds to the yet unidentified Nucleoprotein Receptor, NR, expressed at the surface of human thymic epithelial cells [80].

The interaction between N_{TAIL} and hsp72 stimulates both transcription and genome replication. Two binding sites for hsp72 have been identified [141, 142]. High affinity binding is supported by the α-MORE, and hsp72 can competitively inhibit binding of XD to N_{TAIL} [141]. A second low affinity binding site is present in the C-terminus of N_{TAIL} [142]. Variability in sequence of the N protein C-terminus gives rise to hsp72 binding and non-binding variants. Analysis of infectious virus containing a non-binding motif shows loss of hsp72-dependent stimulation of transcription but not genome replication [141]. These findings suggest two mechanisms by which hsp72 could enhance transcription and genome replication, and both involve reducing the stability of P/N_{TAIL} complexes, thereby promoting successive cycles of binding and release that are essential to polymerase processivity [18, 141]. The first mechanism is competition between hsp72 and XD for α-MORE binding, and this would occur at low hsp72 concentrations. In the second mechanism, hsp72 would neutralize the contribution of the C-terminus of N_{TAIL} to the formation of a stable P-N_{TAIL} complex, and this would occur in the context of elevated cellular levels of hsp72 and only for MeV strains that support hsp72 binding in this region [141]. The basis for the separable effects of hsp72 on genome replication *versus* transcription remains to be shown, with template changes unique to a replicase *versus* transcriptase being a primary candidate. The latter could involve unique nucleocapsid ultrastructural morphologies, with hsp72-dependent morphologies being well-documented for CDV (see [95, 96] and Chapter 3).

As for the functional role of hsp72 in the context of MeV infection, it has been proposed that the elevation in the hsp72 levels in response to the infection could contribute to virus clearance [23, 98]. Indeed, the stimulation of viral transcription and replication by hsp72 is also associated to cytopathic effects leading to apoptosis and release of viral proteins in the extracellular compartment [97, 132, 133]. These would stimulate the adaptative immune response, thereby leading to virus clearance (see Chapter 3).

Additional binding of cellular partners by N_{TAIL} has the potential to influence both innate and adaptive immunity. N_{TAIL} is involved in the interaction with IRF-3. This interaction triggers the phosphorylation-dependent activation of IRF-3 and its consequent nuclear import, followed by induction of pro-inflammatory cytokines [121] (see Chapter 4).

Finally, after apoptosis of infected cells, the viral nucleocapsid is released in the extracellular compartment where it becomes available to cell surface receptors. While N_{CORE} specifically interacts with FcγRII [79], N_{TAIL} interacts with the yet uncharacterized Nucleoprotein Receptor, NR. This latter is expressed at the surface of dendritic cells of lymphoid origin (both normal and tumoral) [80], and of T and B lymphocytes [79]. Flow cytofluorimetry studies carried out on truncated forms of N_{TAIL} allowed the identification of the N_{TAIL} region responsible for the interaction with NR (Box1, aa 401-420) [79]. The N_{TAIL}-NR interaction triggers an arrest in the G_0/G_1 phase of cell cycle whereas the N_{CORE}-FcγRII interaction triggers apoptosis [79]. Both mechanisms have the potential to contribute to immunosuppression that is a hallmark of MeV infections (see [79] and Chapter 5).

SDSL EPR studies in the presence of TFE pointed out a previously undetected structural propensity within Box1 [89]. The biological relevance of the increased rigidity of Box1, may reside in a gain of structure possibly arising upon binding to NR. Indeed, while Box1 is not involved in the interaction with XD [18], it mediates the N_{TAIL}-NR interaction [79]. We can thus speculate that Box1 could undergo induced folding upon binding to NR. However, definitive answers on gain of regular secondary structure elements by Box1 upon binding to NR, awaits the isolation of this receptor and the molecular characterization of its interaction with N_{TAIL}.

The following chapters are devoted to a state-of-the-art on the functional relevance of the interaction between N_{TAIL} and its various intracellular and extracellular partners.

Acknowledgements

We wish to thank all the persons who contributed to the works herein described. In particular, within the AFMB laboratory, we would like to thank Véronique Receveur-Brechot, Hervé Darbon, Benjamin Morin, Bruno Canard, Kenth Johansson, David Karlin, François Ferron, Valérie Campanacci and Christian Cambillau,. We also thank Keith Dunker (Indiana University, USA), David Bhella (MRC, Glasgow, UK), Michael Oglesbee (Ohio State University, USA), Hélène Valentin and Chantal Rabourdin-Combe (INSERM, Lyon, France), André Fournel, Valérie Belle and Bruno Guigliarelli (Bioénergetique et Ingéniere des Proteins, CNRS, Marseille, France). We are grateful to Tim Vojt, David Bhella and David Karlin who are the authors of Figures 8, 10 and 11, respectively. We also want to thank Denis Gerlier for stimulating discussions and for helpful comments on the manuscript. The studies mentioned in this chapter were carried out with the financial support of the European Commission, program RTD, QLK2-CT2001-01225, "Towards the design of new potent antiviral drugs: structure-function analysis of *Paramyxoviridae* polymerase", and of the Agence Nationale de la Recherche, specific program "Microbiologie et Immunologie", ANR-05-MIIM-035-02, "Structure and disorder of measles virus nucleoprotein: molecular partnership and functional impact".

References

[1] Albertini, A. A., A. K. Wernimont, T. Muziol, R. B. Ravelli, C. R. Clapier, G. Schoehn, W. Weissenhorn, and R. W. Ruigrok. 2006. Crystal structure of the rabies virus nucleoprotein-RNA complex. *Science* 313:360-3.

[2] Albertini, A. A. V., G. Schoehn, and R. W. Ruigrok. 2005. Structures impliquées dans la réplication et la transcription des virus à ARN non segmentés de sens négatif. *Virologie* 9:83-92.

[3] Anfinsen, C. B. 1973. Principles that govern the folding of protein chains. *Science* 181:223-30.

[4] Baker, S. C., and S. A. Moyer. 1988. Encapsidation of Sendai virus genome RNAs by purified NP protein during in vitro replication. *J. Virol.* 62:834-8.

[5] Bankamp, B., S. M. Horikami, P. D. Thompson, M. Huber, M. Billeter, and S. A. Moyer. 1996. Domains of the measles virus N protein required for binding to P protein and self-assembly. *Virology* 216:272-7.

[6] Bernado, P., L. Blanchard, P. Timmins, D. Marion, R. W. Ruigrok, and M. Blackledge. 2005. A structural model for unfolded proteins from residual dipolar couplings and small-angle x-ray scattering. *Proc. Natl. Acad. Sci. USA* 102:17002-7.

[7] Bhella, D., A. Ralph, L. B. Murphy, and R. P. Yeo. 2002. Significant differences in nucleocapsid morphology within the Paramyxoviridae. *J. Gen. Virol.* 83:1831-9.

[8] Bhella, D., A. Ralph, and R. P. Yeo. 2004. Conformational flexibility in recombinant measles virus nucleocapsids visualised by cryo-negative stain electron microscopy and real-space helical reconstruction. *J. Mol. Biol.* 340:319-31.

[9] Biswas, R., H. Kuhne, G. W. Brudvig, and V. Gopalan. 2001. Use of EPR spectroscopy to study macromolecular structure and function. *Sci. Prog.* 84:45-67.

[10] Bitko, V., and S. Barik. 2001. Phenotypic silencing of cytoplasmic genes using sequence-specific double-stranded short interfering RNA and its application in the reverse genetics of wild type negative-strand RNA viruses. *BMC Microbiol.* 1:34.

[11] Blanchard, L., N. Tarbouriech, M. Blackledge, P. Timmins, W. P. Burmeister, R. W. Ruigrok, and D. Marion. 2004. Structure and dynamics of the nucleocapsid-binding domain of the Sendai virus phosphoprotein in solution. *Virology* 319:201-11.

[12] Blumberg, B. M., C. Giorgi, and D. Kolakofsky. 1983. N protein of vesicular stomatitis virus selectively encapsidates leader RNA in vitro. *Cell* 32:559-67.

[13] Bogatyreva, N. S., A. V. Finkelstein, and O. V. Galzitskaya. 2006. Trend of amino acid composition of proteins of different taxa. *Journal of Bioinformatics and Computational Biology,* In press.

[14] Bourhis, J., B. Canard, and S. Longhi. 2007. Predicting protein disorder and induced folding: from theoretical principles to practical applications. *Current Protein and Peptide Science* in press.

[15] Bourhis, J., K. Johansson, V. Receveur-Bréchot, C. J. Oldfield, A. K. Dunker, B. Canard, and S. Longhi. 2004. The C-terminal domain of measles virus nucleoprotein belongs to the classof intrinsically disordered proteins that fold upon binding to their pohysiological partner. *Virus Research* 99:157-67.

[16] Bourhis, J. M., B. Canard, and S. Longhi. 2005. Désordre structural au sein du complexe réplicatif du virus de la rougeole: implications fonctionnelles. *Virologie* 9:367-83.

[17] Bourhis, J. M., B. Canard, and S. Longhi. 2006. Structural disorder within the replicative complex of measles virus: functional implications. *Virology* 344:94-110.

[18] Bourhis, J. M., V. Receveur-Bréchot, M. Oglesbee, X. Zhang, M. Buccellato, H. Darbon, B. Canard, S. Finet, and S. Longhi. 2005. The intrinsically disordered C-terminal domain of the measles virus nucleoprotein interacts with the C-terminal domain of the phosphoprotein via two distinct sites and remains predominantly unfolded. *Protein Sci.* 14:1975-1992.

[19] Bowman, P., C. A. Galea, E. Lacy, and R. W. Kriwacki. 2006. Thermodynamic characterization of interactions between p27(Kip1) and activated and non-activated Cdk2: intrinsically unstructured proteins as thermodynamic tethers. *Biochim. Biophys. Acta* 1764:182-9.

[20] Breidenbach, M. A., and A. T. Brunger. 2004. Substrate recognition strategy for botulinum neurotoxin serotype A. *Nature* 432:925-9.

[21] Buchholz, C. J., C. Retzler, H. E. Homann, and W. J. Neubert. 1994. The carboxy-terminal domain of Sendai virus nucleocapsid protein is involved in complex formation between phosphoprotein and nucleocapsid- like particles. *Virology* 204:770-6.

[22] Buchholz, C. J., D. Spehner, R. Drillien, W. J. Neubert, and H. E. Homann. 1993. The conserved N-terminal region of Sendai virus nucleocapsid protein NP is required for nucleocapsid assembly. *J. Virol.* 67:5803-12.

[23] Carsillo, T., M. Carsillo, S. Niewiesk, D. Vasconcelos, and M. Oglesbee. 2004. Hyperthermic pre-conditioning promotes measles virus clearance from brain in a mouse model of persistent infection. *Brain Res.* 1004:73-82.

[24] Cattaneo, R., A. Schmid, P. Spielhofer, K. Kaelin, K. Baczko, V. ter Meulen, J. Pardowitz, S. Flanagan, B. K. Rima, S. A. Udem, and et al. 1989. Mutated and hypermutated genes of persistent measles viruses which caused lethal human brain diseases. *Virology* 173:415-25.

[25] Chen, J. W., P. Romero, V. N. Uversky, and A. K. Dunker. 2006. Conservation of intrinsic disorder in protein domains and families: I. A database of conserved predicted disordered regions. *J. Proteome Res.* 5:879-87.

[26] Chen, J. W., P. Romero, V. N. Uversky, and A. K. Dunker. 2006. Conservation of intrinsic disorder in protein domains and families: II. functions of conserved disorder. *J. Proteome Res.* 5:888-98.

[27] Chen, M., J. C. Cortay, and D. Gerlier. 2003. Measles virus protein interactions in yeast: new findings and caveats. *Virus Res.* 98:123-9.

[28] Cheng, Y., T. Legall, C. J. Oldfield, J. P. Mueller, Y. Y. Van, P. Romero, M. S. Cortese, V. N. Uversky, and A. K. Dunker. 2006. Rational drug design via intrinsically disordered protein. *Trends Biotechnol.* 24:435-42.

[29] Coronel, E. C., T. Takimoto, K. G. Murti, N. Varich, and A. Portner. 2001. Nucleocapsid incorporation into parainfluenza virus is regulated by specific interaction with matrix protein. *J. Virol.* 75:1117-23.

[30] Curran, J., H. Homann, C. Buchholz, S. Rochat, W. Neubert, and D. Kolakofsky. 1993. The hypervariable C-terminal tail of the Sendai paramyxovirus nucleocapsid protein is required for template function but not for RNA encapsidation. *J. Virol.* 67:4358-64.

[31] Curran, J., and D. Kolakofsky. 1999. Replication of paramyxoviruses. *Adv. Virus Res.* 54:403-22.

[32] Curran, J., J. B. Marq, and D. Kolakofsky. 1995. An N-terminal domain of the Sendai paramyxovirus P protein acts as a chaperone for the NP protein during the nascent chain assembly step of genome replication. *J. Virol.* 69:849-55.

[33] Curran, J., T. Pelet, and D. Kolakofsky. 1994. An acidic activation-like domain of the Sendai virus P protein is required for RNA synthesis and encapsidation. *Virology* 202:875-84.

[34] Curran, J. A., and D. Kolakofsky. 1991. Rescue of a Sendai virus DI genome by other parainfluenza viruses: implications for genome replication. *Virology* 182:168-76.

[35] De, B. P., and A. K. Banerjee. 1999. Involvement of actin microfilaments in the transcription/replication of human parainfluenza virus type 3: possible role of actin in other viruses. *Microsc. Res. Tech.* 47:114-23.

[36] Dosztanyi, Z., V. Csizmok, P. Tompa, and I. Simon. 2005. IUPred: web server for the prediction of intrinsically unstructured regions of proteins based on estimated energy content. *Bioinformatics* 21:3433-4.

[37] Dosztanyi, Z., V. Csizmok, P. Tompa, and I. Simon. 2005. The pairwise energy content estimated from amino acid composition discriminates between folded and intrinsically unstructured proteins. *J. Mol. Biol.* 347:827-39.

[38] Dunker, A. K., C. J. Brown, J. D. Lawson, L. M. Iakoucheva, and Z. Obradovic. 2002. Intrinsic disorder and protein function. *Biochemistry* 41:6573-82.

[39] Dunker, A. K., M. S. Cortese, P. Romero, L. M. Iakoucheva, and V. N. Uversky. 2005. Flexible nets. *Febs. J.* 272:5129-5148.

[40] Dunker, A. K., E. Garner, S. Guilliot, P. Romero, K. Albrecht, J. Hart, Z. Obradovic, C. Kissinger, and J. E. Villafranca. 1998. Protein disorder and the evolution of molecular recognition: theory, predictions and observations. *Pac. Symp. Biocomput.* 3:473-84.

[41] Dunker, A. K., J. D. Lawson, C. J. Brown, R. M. Williams, P. Romero, J. S. Oh, C. J. Oldfield, A. M. Campen, C. M. Ratliff, K. W. Hipps, J. Ausio, M. S. Nissen, R. Reeves, C. Kang, C. R. Kissinger, R. W. Bailey, M. D. Griswold, W. Chiu, E. C. Garner, and Z. Obradovic. 2001. Intrinsically disordered protein. *J. Mol. Graph. Model.* 19:26-59.

[42] Dunker, A. K., and Z. Obradovic. 2001. The protein trinity--linking function and disorder. *Nat. Biotechnol.* 19:805-6.

[43] Dyson, H. J., and P. E. Wright. 2002. Coupling of folding and binding for unstructured proteins. *Curr. Opin. Struct. Biol.* 12:54-60.

[44] Dyson, H. J., and P. E. Wright. 2005. Intrinsically unstructured proteins and their functions. *Nat. Rev. Mol. Cell Biol.* 6:197-208.

[45] Egelman, E. H., S. S. Wu, M. Amrein, A. Portner, and G. Murti. 1989. The Sendai virus nucleocapsid exists in at least four different helical states. *J. Virol.* 63:2233-43.

[46] Fearns, R., and P. L. Collins. 1999. Role of the M2-1 transcription antitermination protein of respiratory syncytial virus in sequential transcription. *J. Virol.* 73:5852-64.

[47] Feix, J. B., and C. S. Klug. 1998. Site-directed spin-labeling of membrane proteins and peptide-membrane interactions., p. 251-81. *In* L. Berliner (ed.), In *Biological magnetic resonance*, vol. Spin labeling: the next millenium. Plenum Press, New York.

[48] Ferron, F., S. Longhi, B. Canard, and D. Karlin. 2006. A practical overview of protein disorder prediction methods. *Proteins* 65:1-14.

[49] Ferron, F., S. Longhi, B. Henrissat, and B. Canard. 2002. Viral RNA-polymerases - a predicted 2'-O-ribose methyltransferase domain shared by all Mononegavirales. *Trends Biochem. Sci.* 27:222-4.

[50] Ferron, F., C. Rancurel, S. Longhi, C. Cambillau, B. Henrissat, and B. Canard. 2005. VaZyMolO: a tool to define and classify modularity in viral proteins. *J. Gen. Virol.* 86:743-9.

[51] Fink, A. L. 2005. Natively unfolded proteins. *Curr. Opin. Struct. Biol.* 15:35-41.

[52] Flamand, A., H. Raux, Y. Gaudin, and R. W. Ruigrok. 1993. Mechanisms of rabies virus neutralization. *Virology* 194:302-13.

[53] Fuxreiter, M., I. Simon, P. Friedrich, and P. Tompa. 2004. Preformed structural elements feature in partner recognition by intrinsically unstructured proteins. *J. Mol. Biol.* 338:1015-26.

[54] Garner, E., P. Romero, A. K. Dunker, C. Brown, and Z. Obradovic. 1999. Predicting Binding Regions within Disordered Proteins. *Genome Inform Ser Workshop Genome Inform* 10:41-50.

[55] Giraudon, P., M. F. Jacquier, and T. F. Wild. 1988. Antigenic analysis of African measles virus field isolates: identification and localisation of one conserved and two variable epitope sites on the NP protein. *Virus Res.* 10:137-52.

[56] Gombart, A. F., A. Hirano, and T. C. Wong. 1993. Conformational maturation of measles virus nucleocapsid protein. *J. Virol.* 67:4133-41.

[57] Green, T. J., X. Zhang, G. W. Wertz, and M. Luo. 2006. Structure of the vesicular stomatitis virus nucleoprotein-RNA complex. *Science* 313:357-60.

[58] Gunasekaran, K., C. J. Tsai, S. Kumar, D. Zanuy, and R. Nussinov. 2003. Extended disordered proteins: targeting function with less scaffold. *Trends Biochem. Sci.* 28:81-5.

[59] Gupta, A. K., D. Shaji, and A. K. Banerjee. 2003. Identification of a novel tripartite complex involved in replication of vesicular stomatitis virus genome RNA. *J. Virol.* 77:732-8.

[60] Hartlieb, B., J. Modrof, E. Muhlberger, H. D. Klenk, and S. Becker. 2003. Oligomerization of Ebola virus VP30 is essential for viral transcription and can be inhibited by a synthetic peptide. *J. Biol. Chem.* 278:41830-6.

[61] Harty, R. N., and P. Palese. 1995. Measles virus phosphoprotein (P) requires the NH2- and COOH-terminal domains for interactions with the nucleoprotein (N) but only the COOH terminus for interactions with itself. *J. Gen. Virol.* 76:2863-7.

[62] Heggeness, M. H., A. Scheid, and P. W. Choppin. 1980. Conformation of the helical nucleocapsids of paramyxoviruses and vesicular stomatitis virus: reversible coiling and uncoiling induced by changes in salt concentration. *Proc. Natl. Acad. Sci. USA* 77:2631-5.

[63] Heggeness, M. H., A. Scheid, and P. W. Choppin. 1981. The relationship of conformational changes in the Sendai virus nucleocapsid to proteolytic cleavage of the NP polypeptide. *Virology* 114:555-62.

[64] Horikami, S. M., and S. A. Moyer. 1995. Structure, transcription, and replication of measles virus. *Curr. Top Microbiol. Immunol.* 191:35-50.

[65] Hubbell, W. L., A. Gross, R. Langen, and M. A. Lietzow. 1998. Recent advances in site-directed spin labeling of proteins. *Curr. Opin. Struct. Biol.* 8:649-56.

[66] Huber, M., R. Cattaneo, P. Spielhofer, C. Orvell, E. Norrby, M. Messerli, J. C. Perriard, and M. A. Billeter. 1991. Measles virus phosphoprotein retains the nucleocapsid protein in the cytoplasm. *Virology* 185:299-308.

[67] Iakoucheva, L. M., C. J. Brown, J. D. Lawson, Z. Obradovic, and A. K. Dunker. 2002. Intrinsic disorder in cell-signaling and cancer-associated proteins. *J. Mol. Biol.* 323:573-84.

[68] Iseni, F., F. Baudin, D. Blondel, and R. W. Ruigrok. 2000. Structure of the RNA inside the vesicular stomatitis virus nucleocapsid. *Rna* 6:270-81.

[69] Johansson, K., J. M. Bourhis, V. Campanacci, C. Cambillau, B. Canard, and S. Longhi. 2003. Crystal structure of the measles virus phosphoprotein domain responsible for the induced folding of the C-terminal domain of the nucleoprotein. *J. Biol. Chem.* 278:44567-73.

[70] Karlin, D., F. Ferron, B. Canard, and S. Longhi. 2003. Structural disorder and modular organization in Paramyxovirinae N and P. *J. Gen. Virol.* 84:3239-52.

[71] Karlin, D., S. Longhi, and B. Canard. 2002. Substitution of two residues in the measles virus nucleoprotein results in an impaired self-association. *Virology* 302:420-32.

[72] Karlin, D., S. Longhi, V. Receveur, and B. Canard. 2002. The N-terminal domain of the phosphoprotein of morbilliviruses belongs to the natively unfolded class of proteins. *Virology* 296:251-62.

[73] Kingston, R. L., D. J. Hamel, L. S. Gay, F. W. Dahlquist, and B. W. Matthews. 2004. Structural basis for the attachment of a paramyxoviral polymerase to its template. *Proc. Natl. Acad. Sci. USA* 101:8301-6.

[74] Kingston, R. L., A. B. Walter, and L. S. Gay. 2004. Characterization of nucleocapsid binding by the measles and the mumps virus phosphoprotein. *J. Virol.* 78:8615-29.

[75] Kolakofsky, D., L. Roux, D. Garcin, and R. W. Ruigrok. 2005. Paramyxovirus mRNA editing, the "rule of six" and error catastrophe: a hypothesis. *J. Gen. Virol.* 86:1869-77.

[76] Kouznetzoff, A., M. Buckle, and N. Tordo. 1998. Identification of a region of the rabies virus N protein involved in direct binding to the viral RNA. *J. Gen. Virol.* 79:1005-13.

[77] Kriwacki, R. W., L. Hengst, L. Tennant, S. I. Reed, and P. E. Wright. 1996. Structural studies of p21Waf1/Cip1/Sdi1 in the free and Cdk2-bound state: conformational disorder mediates binding diversity. *Proc. Natl. Acad. Sci. USA* 93:11504-9.

[78] Lacy, E. R., I. Filippov, W. S. Lewis, S. Otieno, L. Xiao, S. Weiss, L. Hengst, and R. W. Kriwacki. 2004. p27 binds cyclin-CDK complexes through a sequential mechanism involving binding-induced protein folding. *Nat. Struct. Mol. Biol.* 11:358-64.

[79] Laine, D., J. Bourhis, S. Longhi, M. Flacher, L. Cassard, B. Canard, C. Sautès-Fridman, C. Rabourdin-Combe, and H. Valentin. 2005. Measles virus nucleoprotein induces cell

proliferation arrest and apoptosis through NTAIL/NR and NCORE/FcgRIIB1 interactions, respectively. *J. Gen. Virol.* 86:1771-84.
[80] Laine, D., M. Trescol-Biémont, S. Longhi, G. Libeau, J. Marie, P. Vidalain, O. Azocar, A. Diallo, B. Canard, C. Rabourdin-Combe, and H. Valentin. 2003. Measles virus nucleoprotein binds to a novel cell surface receptor distinct from FcgRII via its C-terminal domain: role in MV-induced immunosuppression. *J. Virol.* 77:11332-46.
[81] Lamb, R. A., and D. Kolakofsky. 2001. *Paramyxoviridae* : The Viruses and Their Replication. In "*Fields Virology*" (B.N. Fields, D.M. Knipe, and P.M. Howley, Eds.), 4th ed., pp 1305-1340. Lippincott-Raven, Philadelphia, PA.
[82] Li, J., J. T. Wang, and S. P. Whelan. 2006. A unique strategy for mRNA cap methylation used by vesicular stomatitis virus. *Proc. Natl. Acad. Sci. USA* 103:8493-8.
[83] Liston, P., R. Batal, C. DiFlumeri, and D. J. Briedis. 1997. Protein interaction domains of the measles virus nucleocapsid protein (NP). *Arch. Virol.* 142:305-21.
[84] Longhi, S., and B. Canard. 1999. Mécanismes de transcription et de réplication des *Paramyxoviridae. Virologie* 3:227-240.
[85] Longhi, S., V. Receveur-Brechot, D. Karlin, K. Johansson, H. Darbon, D. Bhella, R. Yeo, S. Finet, and B. Canard. 2003. The C-terminal domain of the measles virus nucleoprotein is intrinsically disordered and folds upon binding to the C-terminal moiety of the phosphoprotein. *J. Biol. Chem.* 278:18638-48.
[86] Marion, D., N. Tarbouriech, R. W. Ruigrok, W. P. Burmeister, and L. Blanchard. 2001. Assignment of the 1H, 15N and 13C resonances of the nucleocapsid- binding domain of the Sendai virus phosphoprotein. *J. Biomol. NMR* 21:75-6.
[87] Mavrakis, M., L. Kolesnikova, G. Schoehn, S. Becker, and R. W. Ruigrok. 2002. Morphology of Marburg virus NP-RNA. *Virology* 296:300-7.
[88] McHaourab, H. S., M. A. Lietzow, K. Hideg, and W. L. Hubbell. 1996. Motion of spin-labeled side chains in T4 lysozyme. Correlation with protein structure and dynamics. *Biochemistry* 35:7692-704.
[89] Morin, B., J. M. Bourhis, V. Belle, M. Woudstra, F. Carrière, B. BGuigliarelli, A. Fournel, and S. Longhi. 2006. Assessing induced folding of an intrinsically diosrdred protein by site-directed spin-labeling EPR spectroscopy. *J. Phys. Chem.* B 110:20596-608.
[90] Moyer, S. A., S. C. Baker, and S. M. Horikami. 1990. Host cell proteins required for measles virus reproduction. *J. Gen. Virol.* 71:775-83.
[91] Myers, T. M., A. Pieters, and S. A. Moyer. 1997. A highly conserved region of the Sendai virus nucleocapsid protein contributes to the NP-NP binding domain. *Virology* 229:322-35.
[92] Myers, T. M., S. Smallwood, and S. A. Moyer. 1999. Identification of nucleocapsid protein residues required for Sendai virus nucleocapsid formation and genome replication. *J. Gen. Virol.* 80:1383-91.
[93] Nishio, M., M. Tsurudome, M. Ito, M. Kawano, S. Kusagawa, H. Komada, and Y. Ito. 1999. Mapping of domains on the human parainfluenza virus type 2 nucleocapsid protein (NP) required for NP-phosphoprotein or NP-NP interaction. *J. Gen. Virol.* 80:2017-22.

[94] Ogino, T., M. Kobayashi, M. Iwama, and K. Mizumoto. 2005. Sendai virus RNA-dependent RNA polymerase L protein catalyzes cap methylation of virus-specific mRNA. *J. Biol. Chem.* 280:4429-35.

[95] Oglesbee, M., S. Ringler, and S. Krakowka. 1990. Interaction of canine distemper virus nucleocapsid variants with 70K heat-shock proteins. *J. Gen. Virol.* 71:1585-90.

[96] Oglesbee, M., L. Tatalick, J. Rice, and S. Krakowka. 1989. Isolation and characterization of canine distemper virus nucleocapsid variants. *J. Gen. Virol.* 70 (Pt 9):2409-19.

[97] Oglesbee, M. J., H. Kenney, T. Kenney, and S. Krakowka. 1993. Enhanced production of morbillivirus gene-specific RNAs following induction of the cellular stress response in stable persistent infection. *Virology* 192:556-67.

[98] Oglesbee, M. J., M. Pratt, and T. Carsillo. 2002. Role for heat shock proteins in the immune response to measles virus infection. *Viral. Immunol.* 15:399-416.

[99] Oldfield, C. J., Y. Cheng, M. S. Cortese, C. J. Brown, V. N. Uversky, and A. K. Dunker. 2005. Comparing and Combining Predictors of Mostly Disordered Proteins. *Biochemistry* 44:1989-2000.

[100] Oldfield, C. J., Y. Cheng, M. S. Cortese, P. Romero, V. N. Uversky, and A. K. Dunker. 2005. Coupled Folding and Binding with alpha-Helix-Forming Molecular Recognition Elements. *Biochemistry* 44:12454-12470.

[101] Plumet, S., W. P. Duprex, and D. Gerlier. 2005. Dynamics of viral RNA synthesis during measles virus infection. *J. Virol.* 79:6900-8.

[102] Precious, B., D. F. Young, A. Bermingham, R. Fearns, M. Ryan, and R. E. Randall. 1995. Inducible expression of the P, V, and NP genes of the paramyxovirus simian virus 5 in cell lines and an examination of NP-P and NP-V interactions. *J. Virol.* 69:8001-10.

[103] Rahaman, A., N. Srinivasan, N. Shamala, and M. S. Shaila. 2004. Phosphoprotein of the rinderpest virus forms a tetramer through a coiled coil region important for biological function. A structural insight. *J. Biol. Chem.* 279:23606-14.

[104] Receveur-Bréchot, V., J. M. Bourhis, V. N. Uversky, B. Canard, and S. Longhi. 2006. Assessing protein disorder and induced folding. Proteins: Structure, Function and *Bioinformatics* 62:24-45.

[105] Robbins, S. J., and R. H. Bussell. 1979. Structural phosphoproteins associated with purified measles virions and cytoplasmic nucleocapsids. *Intervirology* 12:96-102.

[106] Robbins, S. J., R. H. Bussell, and F. Rapp. 1980. Isolation and partial characterization of two forms of cytoplasmic nucleocapsids from measles virus-infected cells. *J. Gen. Virol.* 47:301-10.

[107] Romero, P., Z. Obradovic, X. Li, E. C. Garner, C. J. Brown, and A. K. Dunker. 2001. Sequence complexity of disordered proteins. *Proteins* 42:38-48.

[108] Roux, L. 2005. Dans le génome des Paramyxovirinae, les promoteurs et leurs activités sont façonnés par la « règle de six. *Virologie* 9:19-34.

[109] Ryan, K. W., and A. Portner. 1990. Separate domains of Sendai virus P protein are required for binding to viral nucleocapsids. *Virology* 174:515-21.

[110] Sato, H., M. Masuda, R. Miura, M. Yoneda, and C. Kai. 2006. Morbillivirus nucleoprotein possesses a novel nuclear localization signal and a CRM1-independent nuclear export signal. *Virology* 352:121-30.

[111] Schoehn, G., F. Iseni, M. Mavrakis, D. Blondel, and R. W. Ruigrok. 2001. Structure of recombinant rabies virus nucleoprotein-RNA complex and identification of the phosphoprotein binding site. *J. Virol.* 75:490-8.

[112] Schoehn, G., M. Mavrakis, A. Albertini, R. Wade, A. Hoenger, and R. W. Ruigrok. 2004. The 12 A structure of trypsin-treated measles virus N-RNA. *J. Mol. Biol.* 339:301-12.

[113] Sivakolundu, S. G., D. Bashford, and R. W. Kriwacki. 2005. Disordered p27Kip1 exhibits intrinsic structure resembling the Cdk2/cyclin A-bound conformation. *J. Mol. Biol.* 353:1118-28.

[114] Smallwood, S., K. W. Ryan, and S. A. Moyer. 1994. Deletion analysis defines a carboxyl-proximal region of Sendai virus P protein that binds to the polymerase L protein. *Virology* 202:154-63.

[115] Spehner, D., R. Drillien, and P. M. Howley. 1997. The assembly of the measles virus nucleoprotein into nucleocapsid-like particles is modulated by the phosphoprotein. *Virology* 232:260-8.

[116] Spehner, D., A. Kirn, and R. Drillien. 1991. Assembly of nucleocapsidlike structures in animal cells infected with a vaccinia virus recombinant encoding the measles virus nucleoprotein. *J. Virol.* 65:6296-300.

[117] Stallcup, K. C., S. L. Wechsler, and B. N. Fields. 1979. Purification of measles virus and characterization of subviral components. *J. Virol.* 30:166-76.

[118] Sweetman, D. A., J. Miskin, and M. D. Baron. 2001. Rinderpest virus C and V proteins interact with the major (L) component of the viral polymerase. *Virology* 281:193-204.

[119] Tapparel, C., D. Maurice, and L. Roux. 1998. The activity of Sendai virus genomic and antigenomic promoters requires a second element past the leader template regions: a motif (GNNNNN)3 is essential for replication. *J. Virol.* 72:3117-28.

[120] Tarbouriech, N., J. Curran, R. W. Ruigrok, and W. P. Burmeister. 2000. Tetrameric coiled coil domain of Sendai virus phosphoprotein. *Nat. Struct. Biol.* 7:777-81.

[121] tenOever, B. R., M. J. Servant, N. Grandvaux, R. Lin, and J. Hiscott. 2002. Recognition of the Measles Virus Nucleocapsid as a Mechanism of IRF-3 Activation. *J. Virol.* 76:3659-69.

[122] Tompa, P. 2002. Intrinsically unstructured proteins. *Trends Biochem Sci*. 27:527.

[123] Tompa, P. 2003. The functional benefits of disorder. *J. Mol. Structure* (Theochem) 666-67:361-71.

[124] Tompa, P. 2005. The interplay between structure and function in intrinsically unstructured proteins. *FEBS Lett.* 579:3346-54.

[125] Tompa, P., Z. Dosztanyi, and I. Simon. 2006. Prevalent structural disorder in E. coli and S. cerevisiae proteomes. *J. Proteome Res*. 5:1996-2000.

[126] Tuckis, J., S. Smallwood, J. A. Feller, and S. A. Moyer. 2002. The C-terminal 88 amino acids of the Sendai virus P protein have multiple functions separable by mutation. *J. Virol.* 76:68-77.

[127] Uversky, V. N. 2002. Natively unfolded proteins: a point where biology waits for physics. *Protein Sci.* 11:739-56.

[128] Uversky, V. N. 2002. What does it mean to be natively unfolded? *Eur. J. Biochem.* 269:2-12.

[129] Uversky, V. N., J. R. Gillespie, and A. L. Fink. 2000. Why are "natively unfolded" proteins unstructured under physiologic conditions? *Proteins* 41:415-427.

[130] Uversky, V. N., J. Li, P. Souillac, R. Jakes, M. Goedert, and A. L. Fink. 2002. Biophysical properties of the synucleins and their propensities to fibrillate: inhibition of alpha-synuclein assembly by beta- and gamma- synucleins. *J. Biol. Chem.* 25:25.

[131] Uversky, V. N., C. J. Oldfield, and A. K. Dunker. 2005. Showing your ID: intrinsic disorder as an ID for recognition, regulation and cell signaling. *J. Mol. Recognit.* 18:343-84.

[132] Vasconcelos, D., E. Norrby, and M. Oglesbee. 1998. The cellular stress response increases measles virus-induced cytopathic effect. *J. Gen. Virol.* 79 (Pt 7):1769-73.

[133] Vasconcelos, D. Y., X. H. Cai, and M. J. Oglesbee. 1998. Constitutive overexpression of the major inducible 70 kDa heat shock protein mediates large plaque formation by measles virus. *J. Gen. Virol.* 79 (Pt 9):2239-47.

[134] Vincent, S., I. Tigaud, H. Schneider, C. J. Buchholz, Y. Yanagi, and D. Gerlier. 2002. Restriction of measles virus RNA synthesis by a mouse host cell line: trans-complementation by polymerase components or a human cellular factor(s). *J. Virol.* 76:6121-30.

[135] Vulliemoz, D., and L. Roux. 2001. "Rule of six": how does the Sendai virus RNA polymerase keep count? *J. Virol.* 75:4506-18.

[136] Warnes, A., A. R. Fooks, A. B. Dowsett, G. W. Wilkinson, and J. R. Stephenson. 1995. Expression of the measles virus nucleoprotein gene in Escherichia coli and assembly of nucleocapsid-like structures. *Gene* 160:173-8.

[137] Williams, R. M., Z. Obradovi, V. Mathura, W. Braun, E. C. Garner, J. Young, S. Takayama, C. J. Brown, and A. K. Dunker. 2001. The protein non-folding problem: amino acid determinants of intrinsic order and disorder. *Pac. Symp. Biocomput.* :89-100.

[138] Wright, P. E., and H. J. Dyson. 1999. Intrinsically unstructured proteins: re-assessing the protein structure-function paradigm. *J. Mol. Biol.* 293:321-31.

[139] Yeung, L. F., P. Lurie, G. Dayan, E. Eduardo, P. H. Britz, S. B. Redd, M. J. Papania, and J. F. Seward. 2005. A limited measles outbreak in a highly vaccinated US boarding school. *Pediatrics* 116:1287-91.

[140] Zandotti, C., D. Jeantet, F. Lambert, D. Waku-Kouomou, F. Wild, F. Freymuth, J. R. Harle, X. de Lamballerie, and R. N. Charrel. 2004. Re-emergence of measles among young adults in Marseilles, France. *Eur. J. Epidemiol.* 19:891-3.

[141] Zhang, X., J. M. Bourhis, S. Longhi, T. Carsillo, M. Buccellato, B. Morin, B. Canard, and M. Oglesbee. 2005. Hsp72 recognizes a P binding motif in the measles virus N protein C-terminus. *Virology* 337:162-74.

[142] Zhang, X., C. Glendening, H. Linke, C. L. Parks, C. Brooks, S. A. Udem, and M. Oglesbee. 2002. Identification and characterization of a regulatory domain on the carboxyl terminus of the measles virus nucleocapsid protein. *J. Virol.* 76:8737-46.

In: Measles Virus Nucleoprotein
Editors: S. Longhi, pp. 37-50
ISBN: 978-1-60021-629-9

Chapter II

Measles Virus Nucleocapsid Structure, Conformational Flexibility and the Rule of Six

David Bhella *
MRC Virology Unit,
Church Street, Glasgow,
G11 5JR UK

Abstract

The intrinsically disordered nature of *Paramyxovirinae* N proteins coupled with the flexibility of N-RNA complexes such as nucleocapsids (NCs), presents significant challenges to their structure analysis. Electron microscopy and image reconstruction has provided a low-resolution insight into NC morphology, highlighting considerable conformational flexibility. NCs range in pitch (the axial rise per helix turn) from 46 to 66 Å and also adopt extended forms with a pitch of 37.5 nm. NCs also vary significantly in twist, with values ranging between at least 12.35 and 13.6 subunits per turn. These variations in conformation appear to be modulated by the natively unfolded C-terminal region, N_{TAIL}, leading to the suggestion that changes in helical parameters may play a role in regulating some aspect of the viral replication cycle, possibly regulating the switch between transcription and replication of the viral genome.

List of Abbreviations

3D	Three-Dimensional
IHRSR	Iterative Helical Real Space Reconstruction
MeV	Measles Virus

* Tel: (44) 141 330 3685; Fax: (44) 141 330 2236; E-mail d.bhella@mrcvu.gla.ac.uk

N	Nucleoprotein
NC	Nucleocapsid
nm	nanometre
P	Phosphoprotein
RNA	Ribonucleic Acid
RSV	Respiratory Syncytial Virus
RV	Rabies Virus
SeV	Sendai Virus
VSV	Vesicular Stomatitis Virus

1. Introduction

Paramyxovirinae nucleoproteins (N) encapsidate their cognate RNA by assembling directly onto the viral genome, forming a helical polymer that is approximately 20 nm in diameter and 1.1 μm long. The helical N-RNA complex or *nucleocapsid* (NC) has a characteristic 'herringbone' morphology when viewed in the transmission electron microscope (Figure 1). Structure analysis of negative sense RNA virus N proteins has proven to be a challenging area of research owing to the intrinsic disorder found in regions of the protein and also to the flexible nature of N multimers such as nucleocapsids. In recent years some progress has been made through the application of a variety of biophysical methods. Spectroscopic approaches combined with bioinformatics analyses have revealed the disordered nature of the C-terminal domain (N_{TAIL}) of Paramyxovirinae N proteins [5, 12, 13] (see also Chapter 1), while crystallography and electron cryomicroscopy are starting to provide the first glimpses of the 3D structure of these proteins.

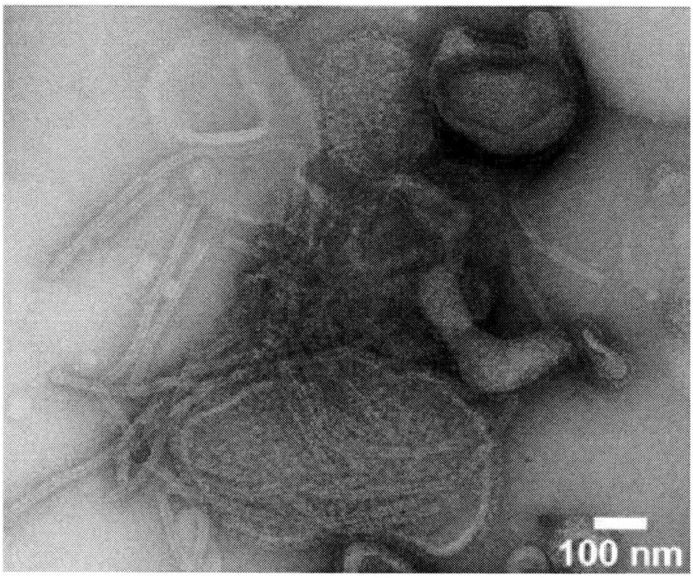

Figure 1. Negative stain transmission electron micrograph of SeV, showing the characteristic herringbone morphology of Paramyxovirinae nucleocapsids.

2. Helical Reconstruction of Sendai Virus Nucleocapsids

The first insights into the three-dimensional (3D) structure of *Paramyxovirinae* NCs came through the application of helical reconstruction methods to transmission electron micrographs of NCs isolated from Sendai virus (SeV) [9]. Calculation of the 3D density was accomplished by Fourier-Bessel analysis and reconstruction [7] of negatively stained images of NCs isolated from purified virions. This image reconstruction method has been widely used in the study of helical macromolecular assemblies but is critically dependent on a high level of order in the filament. Egelman *et al.* noted the presence of several discrete helical conformations, recording pitch measurements of 53 Å and 68 Å in negative stain images as well as an extended 37.5 nm form seen only in metal shadowed preparations, a technique that also revealed that SeV NCs are left-handed helices. Variations in the helical twist were also observed and a single reconstruction was calculated with a pitch of 53 Å and 13.07 subunits per helix turn. A sub-set of NCs with a 53 Å pitch, but a different distribution of helical layer-lines in the computed Fourier transforms were proposed to comprise either 12.78 subunits per turn or 13.22 units. Ambiguities in assigning a handedness to the 13-start helix preclude further certainty concerning these structures.

3D reconstructions reveal that SeV NCs are approximately 20 nm in diameter with a hollow core of around 5 nm diameter, the cavity at the helix centre is however larger at 7.5 nm, but precesses about the helix axis, giving rise to a helical groove along the NC interior. Sections perpendicular to the helix axis revealed that NCs comprise a continuous ring of density 3.5 nm thick with radially protruding spokes of density 2.5 nm long. Sections through the helix axis show the N-monomer density to comprise two globular domains, one contributing to the internal continuous ring, and one at the end of the protruding spoke.

3. Electron Microscopy of Measles Virus N-RNA

Expression of *Paramyxovirinae* nucleoproteins in heterologous systems leads to the non-specific encapsidation of cellular RNA by N and the formation of helical ribonucleoprotein structures, referred to as NC-like particles, that are morphologically indistinguishable from viral NCs [3, 10, 20] (Figure 2A). In addition to helices, ring structures, that may represent single helix turns, are also found in preparations of recombinant N-RNA. Two-dimensional analysis and averaging of rings imaged end-on (top-views) provides a view similar to that of a cross-section perpendicular to the helix axis (Figure 2B). Measles virus (MeV) N-RNA rings comprise 13 N subunits and show a morphology similar to that observed in the SeV NC reconstruction. Two-dimensional analysis of N-RNA rings derived from recombinant expression of related *Paramyxovirinae* nucleoproteins reveals that some viruses may comprise 14 subunits per helix turn, while the more distantly related pneumovirus respiratory syncytial virus (RSV) has 10 or 11 N subunits per helix turn (Figures 2 C and D). Decameric and undecameric rings have also been described as a product of recombinant expression of *Rhabdoviridae* N-proteins. *Rhabdoviridae* are also single-stranded negative-sense RNA viruses, and have much in common with *Paramyxoviridae*. The structures of these N proteins are discussed in section 7.

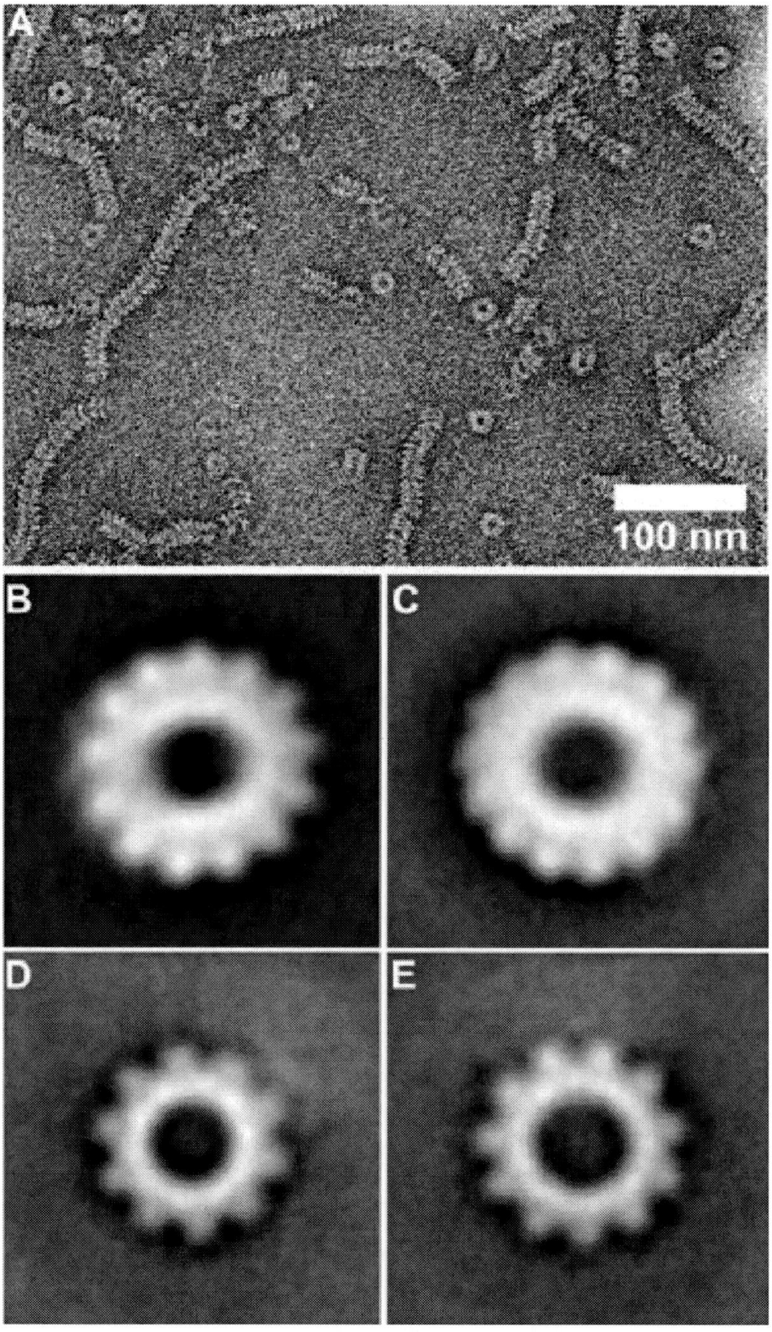

Figure 2. (A) Negative stain transmission electron micrograph of MeV NCs produced by expression of N in insect cells. (B) Two dimensional averages calculated from top-view images of MeV N-RNA rings, showing approximately 13 subunits per helix turn. (C) SV5 N-RNA rings indicate 14 subunits per helix turn while in (D) and (E) RSV N-RNA rings comprise either 10 or 11 N subunits [3].

4. 3D Reconstruction of Nucleocapsids Derived from Recombinantly Expressed Full-Length N

Cryo-negative stained imaging [1] of NC-like structures produced by expression of MeV N protein in insect cells revealed well-preserved helical structures that were nonetheless flexible and difficult to analyse by traditional Fourier-Bessel helical reconstruction methods [4] (Figure 3). An alternative approach to image reconstruction was therefore taken in which short sections of helix were classified according to their helical parameters prior to calculation of 3D volumes using the iterative real space helical reconstruction method (IHRSR) [8].

Sorting was accomplished by cross-correlation of raw images with projections calculated from preliminary reconstructions. Initially data were sorted according to pitch, revealing that values ranged from 50 to 66 Å (Figure 4A). Further analysis of pitch variation indicated that helices varied in pitch along their length, although helices could be described as having an overall longer or shorter pitch. This may explain the apparently discrete pitch measurements for SeV reported by Egelman *et al.* [9] as measurement of layer-spacings in Fourier transforms will give an average pitch measurement over the length of the filament, rather than local pitch estimates.

Classification of helix-sections by twist proved to be a more challenging process. Sorting was accomplished by comparison of raw data with a series of models ranging in twist from 12 to 14 subunits per turn. Models were derived from preliminary reconstructions calculated for each pitch class.

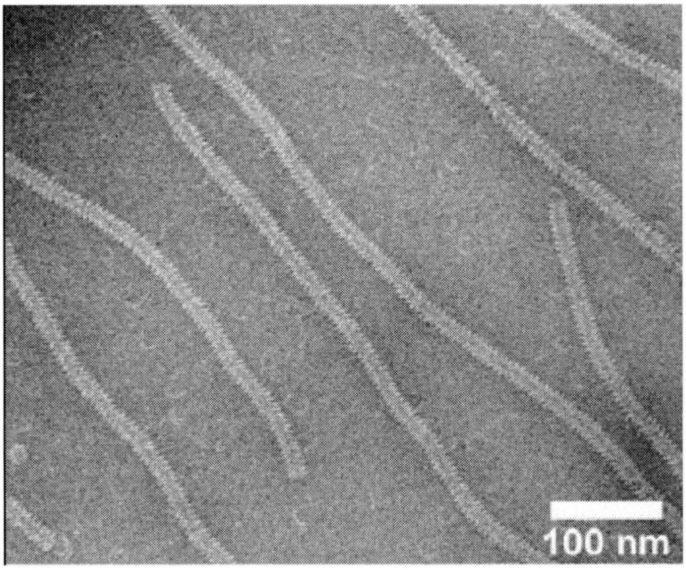

Figure 3. Cryonegative stain electron microscopy of recombinantly expressed MeV NCs, shows superior preservation of NC morphology. Cryogenic methods prevent damage to delicate specimens caused by traditional negative stain methods such as drying or flattening [4].

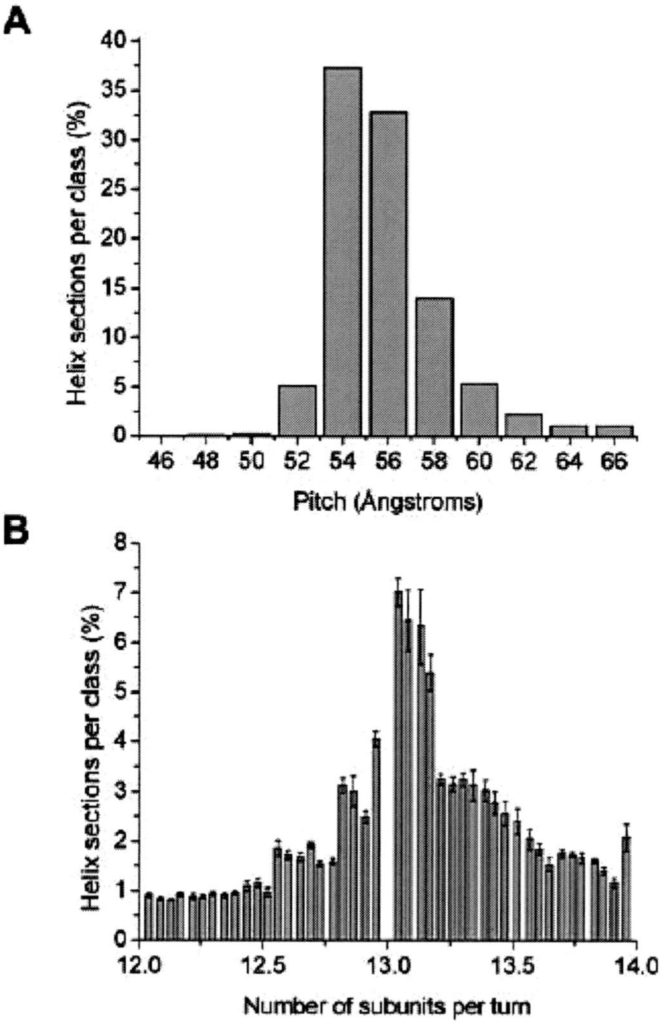

Figure 4. (A) Histogram showing the distribution of pitch values, measured in short sections of MeV NC, imaged by cryo-negative stain, showing that the most populous pitch class is 54 Å. (B) Helix sections were subsequently classified according to twist, showing a broad distribution of values from 12 to 14 subunits per turn in all pitch classes. A histogram of the mean and standard error of the percentage of helix sections in each twist class, for all pitch classes, shows that twist is independent of pitch [4].

These initial reconstructions indicated that the consensus structure for each pitch class comprised between 13.07 and 13.11 subunits per turn, although occasionally stable solutions were found that generated structures with between 12.88 and 12.93 subunits per turn. Such ambiguities are reminiscent of the difficulties previously described by Egelman *et al.* in their investigation of SeV NCs.

Upon sorting, the most populous classes were found to range from 12.8 to 13.6 subunits per turn in all pitch classes (Figure 4B). Reconstructions were calculated for each twist class at every pitch. IHRSR analysis of larger datasets (comprising > 400 helix sections) resulted in the stable calculation of reconstructions that were consistent with the models used in classification. The less populous outlying classes were found to have inconsistent density or did not reach stable solutions through the IHRSR method. As in the preliminary studies, it was observed that many datasets were capable of producing more than one stable solution, such that helix sections classified as comprising 13.2 subunits per turn might also produce a reconstruction with a twist value of 12.8 subunits per turn and vice versa. Based on prior two-dimensional analyses of N-RNA rings (see section 3), reconstructions comprising between 13 and 13.5 subunits per turn were presented (Figure 5).

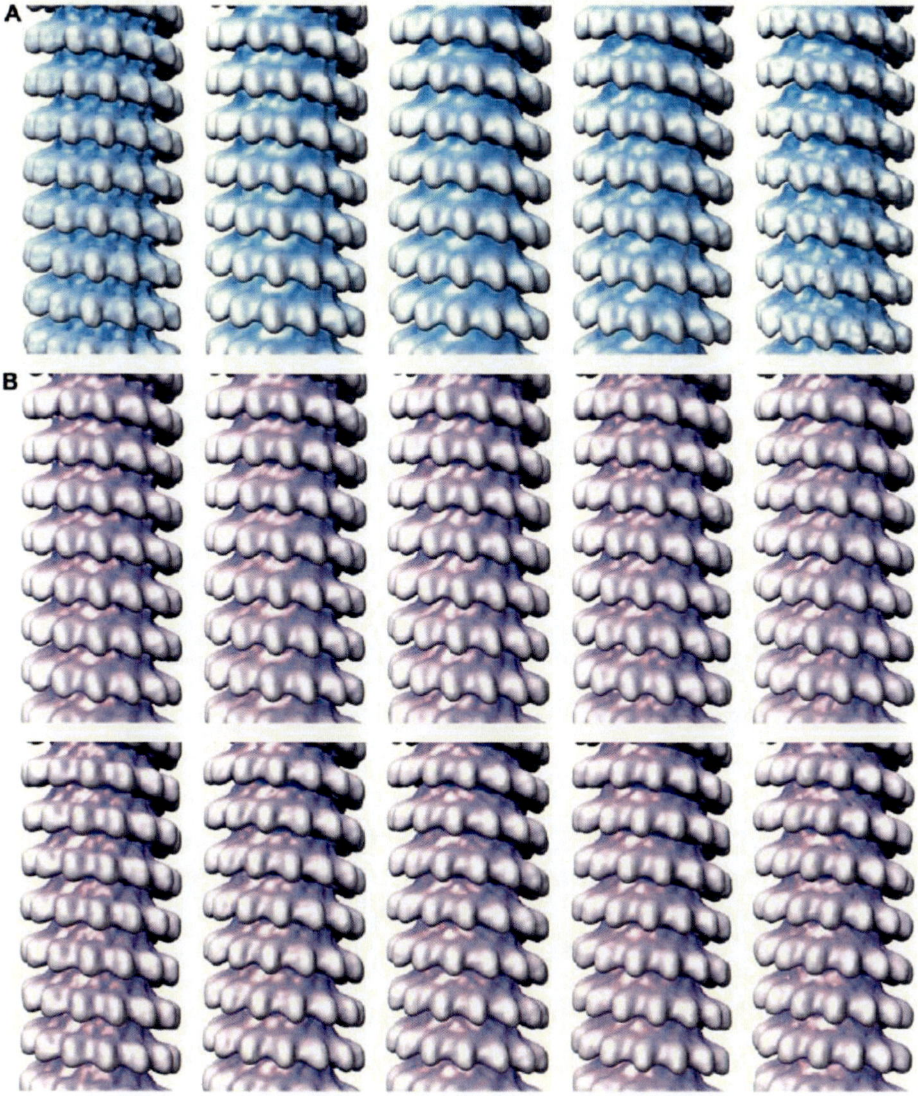

Figure 5. 3D reconstructions of MeV NCs produced by expression of N in insect cells. (A) Surface contoured representations of helices comprising 13.09 subunits per turn and ranging in pitch from 52 to 60 Å (left to right). (B) 54 Å pitch NC reconstructions showing twist variation from 13.04 (top left) to 13.44 (bottom right) subunits per turn [4].

It is entirely possible however that a broader range of conformations may be adopted, as proposed by Schoehn *et al.* in their investigation of trypsin digested MeV N-RNA [18] (see section 5). MeV NC morphology largely resembles that previously described for SeV, although improved resolution of subunit structure was achieved (Figure 6). N comprises two globular domains, one, proximal to the helix axis, which forms a continuous ring of density. Spokes radiate from this core, terminating in the distal, second globular domain. Summing the density for a single turn of the helix by calculating a projection along the helix axis reveals that there is a clear modulation of density along the continuous helix core. Sectioning the NC reconstruction along the helix axis shows a groove of lower density between the two globular regions that was proposed as a possible location of the genomic RNA, providing a good balance between protection and accessibility.

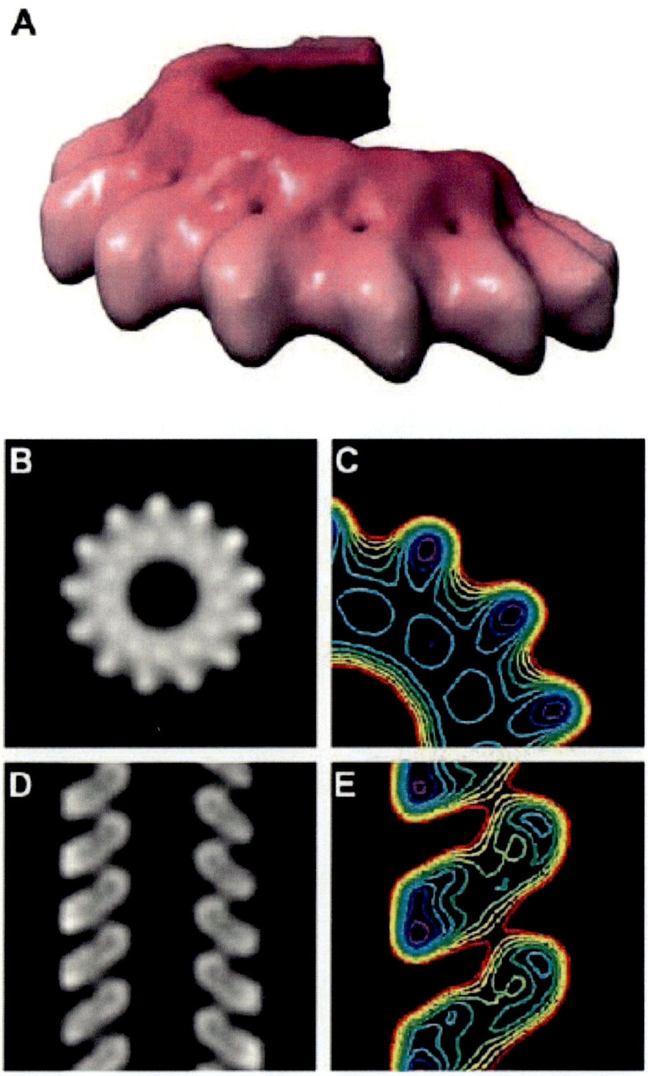

Figure 6. (A) Surface contoured representation of a single helix turn from the 54 Å pitch 13.09 subunit per turn NC reconstruction, showing the density distribution in individual N monomers. Cross sections perpendicular (B,C) and parallel (D, E) t o the helix axis highlight the modulating density in the continuous core region of the NC and the groove running between two globular domains of N. [4].

5. 3D Reconstruction of Trypsin Digested MEV Nucleocapsids

Trypsin treatment of *Paramyxovirinae* NCs to remove the protease-sensitive N_{TAIL} domain results in more rigid helices with a shorter pitch [4, 13, 14, 18]. Schoehn *et al.* exploited this trait to calculate intermediate-resolution (up to 12 Å) reconstructions from cryomicroscopy images of MeV NCs after trypsin treatment. Metal-shadowing experiments confirmed that, like SeV, MeV NCs are left-handed helices and have multiple conformations (Figure 7). Despite the improved order in these structures, Fourier-Bessel methods did not yield higher-resolution reconstructions, IHRSR was therefore used to calculate three reconstructions of helices comprising 11.64, 12.35 and 12.56 subunits per turn and with pitches of 48-50 Å. A model-based classification approach to determine the helical parameters of trypsin digested MeV NCs, imaged by cryo-negative stain, indicated a range of pitch values from 46 to 50 Å. Sorting these data against models comprising between 12 and 14 subunits per turn resulted in several classes with greater numbers of helix sections, suggesting the possibility of discrete twist groups of 12.3, 12.6, 13.1, 13.3 and 13.6 subunits per turn, the most populous class having 13.3 subunits per turn [4]. Again the distribution of twist values hints that these discrete classes may be artefactual, a problem discussed by Egelman (2006).

It is the opinion of the author that although some ambiguity remains over the conformations of trypsin digested MeV NCs, given the higher resolution achieved for reconstructions with 12.35 and 12.56 subunits per turn, by Schoehn *et al.*, these would seem the more likely solutions (Figure 8). In these structures, the N protein has two points of contact with neighbouring molecules, one at the end closest to the helix axis and one approximately halfway along.

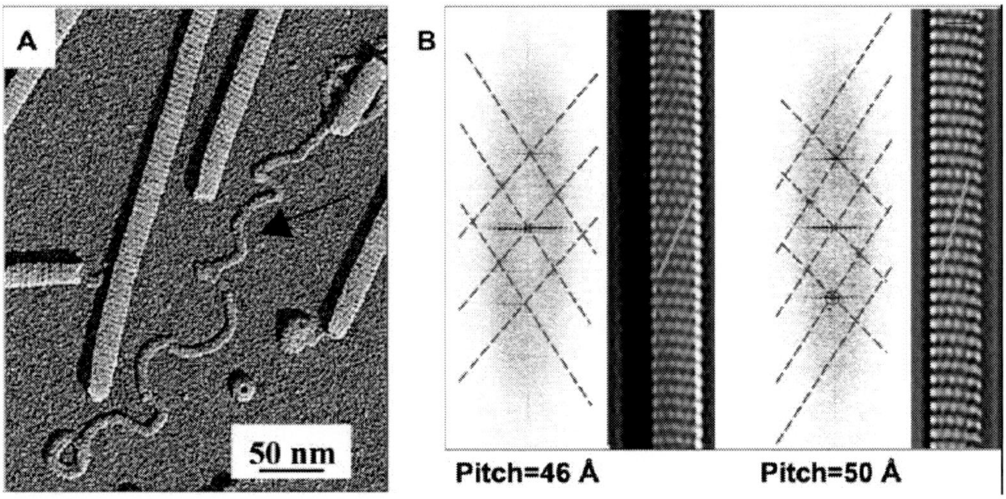

Figure 7. (A) Transmission electron microscopy of metal shadowed trypsin-digested MeV NCs confirms that they are left-handed helices. B. Fourier analysis to calculate helical diffraction patterns from individual NC images shows that they have different pitch and twist values [18].

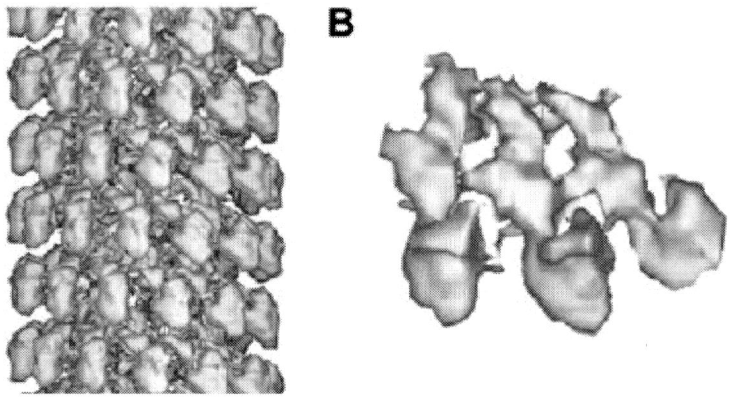

Figure 8. (A) Three-dimensional reconstruction of trypsin-digested MeV NC at 12 Å resolution. The pitch of this reconstruction is 51 Å and it comprises 12.35 subunits per turn. (B) Close up view, perpendicular to the helix axis, of three N-monomers, showing two points of contact between subunits. [18].

There are no contacts between successive turns of the helix. Schoehn *et al.* describe N as consisting of three domains, a globular domain closest to the helix axis, a large distal domain and an elongated stalk that joins the two globular regions.N is 80 Å long and the distal domain is 28 Å wide and 36 Å high. Cis-Platinum labelling of trypsin digested NCs was used to determine the location of RNA within the NC. Experiments showed enhanced density close to the helix axis, but could not definitively identify which of the points of contact between adjacent N-subunits might contain the RNA.

6. Conformational Flexibility and the Rule of Six

Structural investigations of MeV NCs have indicated that these helices are capable of adopting a range of conformations and that removal of the C-terminal N_{TAIL} domain modulates these conformations. The C-terminal region of N has been shown to be critical in mediating many of the important interactions between N and it's co-factors. Moreover N_{TAIL} has been shown to be natively unfolded, undergoing α-helical induced folding upon binding to the 'X-domain' (XD) of the viral P protein [5, 13] (see also Chapter 1), which mediates the interaction between the NC and the viral RNA-dependant RNA polymerase L. Removal of N_{TAIL} results in a shortening of the helix pitch, and changes in twist from approximately 13.1 subunits per turn to either (or all of) 12.3, 12.6, 13.3 or 13.6 subunits per turn. We suggested that the structural transitions induced by removal of N_{TAIL} may mimic changes brought about by interactions between N and its viral or cellular co-factors (4). Preliminary results indicate that XD triggers an unwinding of the MeV NC, thus providing a mechanistic basis to explain enhanced genome accessibility to the polymerase complex as a prerequisite for transcription and replication to occur (Bhella and Longhi, unpublished results).

A

B C

D E

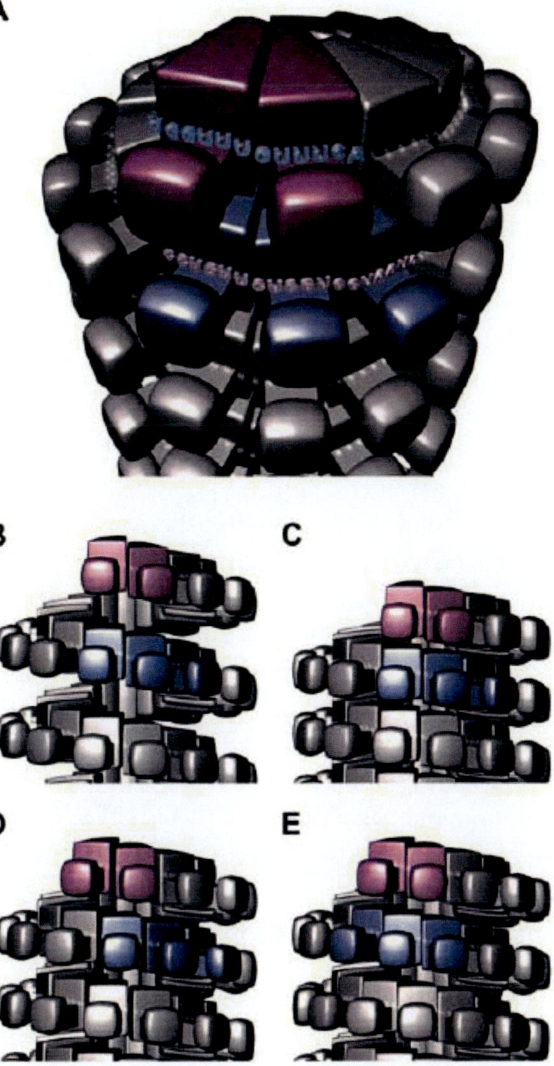

Figure 9. (A) Schematic representation of the MeV genomic promoter. It is composed of two discrete elements that are located on successive turns of the helix, possibly constituting a 'polymerase landing pad'. (B, C) Variation in pitch from 46 to 66 Å, results in significant displacement of promoter elements, as does variation in twist from 12.3 to 13.6 subunits per turn (D, E) We have suggested that such variations in presentation of the promoter may have functional implications, perhaps serving to signal to the polymerase whether it should engage in transcription or replication [4].

Labelling experiments indicate that the viral genome may be located close to the helix axis [18]. The diameter of the hollow core at the centre of these helices is not, however, large enough to accommodate the viral RNA polymerase, which is composed of the ~220 kDa L-protein and four copies of P [22]. The requirement for concurrent synthesis and encapsidation of anti-genomes also strongly argues against the movement of the polymerase along the central cavity. It seems likely then, that conformational changes in the NC will be necessary to allow the polymerase to access genomic RNA. This requirement may account for the influence of N_{TAIL} on helix conformation, as could the need to package NCs within roughly spherical virions. The rule of six (see below), a unique feature of these viruses, raises a third possibility concerning the significance of conformational flexibility in *Paramyxovirinae* NCs, suggesting that changes in helical twist brought about by induced folding of N_{TAIL} could play a regulatory role in the viral replication-cycle.

Egelman *et al.* deduced from their analysis of SeV NCs, that 6 nucleotides of RNA will bind to each N monomer. It has also been established that there is an absolute requirement in both the respiroviruses (the virus family that includes SeV) and morbilliviruses (that which contains MeV), for the viral genome to be of a length that is a multiple of six bases (the rule of six – [6, 16, 19]). The rule of six is not a simple reflection of the 6:1 stoichiometry however. Experiments with mini-genomes have demonstrated that the *Paramyxovirinae* promoter is bi-partite, consisting of two elements whose spatial relationship is critically important [21]. The first element a 12-nucleotide region at the extreme 3' end of the genome is predicted to associate with the first two N subunits of the NC. The second element comprises a triplet repeat of a hexameric motif (3'-CNNNNN-5') between bases 79 and 96 that would bind to the 14th, 15th and 16th N subunits. The two promoter elements therefore are located adjacent to each other on successive turns of the helical NC (Figure 9A) and have been proposed to constitute a 'polymerase landing pad'. The strict requirement imposed on the genome length by this promoter structure is a considerable restriction on mutability in these viruses and must have arisen to fulfil a central role in the replication cycle. It is also apparent that the structural plasticity observed would have a significant impact on the spatial relationship between promoter elements (Figure 9B to E).

This has led to the suggestion that regulated structural changes brought about by induced folding of N_{TAIL} could play a role in regulating some aspect of RNA synthesis by signalling to the polymerase via the relative position of promoter sequences [4]. Central to the viral replication strategy is regulation of the switch between messenger RNA synthesis and antigenome synthesis. The prevailing theory concerning this switch, is that accumulation of soluble N initiates antigenome synthesis by encapsidating nascent RNA and enhancing polymerase processivity [15]. Studies of SeV have indicated that the polymerase is not predetermined to synthesise one or other RNA, rather the mode of synthesis is decided precisely at the point of attachment to the NC [23]. The possibility that synthesis of the appropriate RNA could therefore be regulated by signalling between the NC and polymerase represents an interesting, if difficult to investigate, proposal.

7. N-Protein Structure in Related Viruses

The structures of both Rabies Virus (RV) and Vesicular Stomatitis Virus (VSV) N proteins have been solved by x-ray crystallography revealing the nature of N-N interactions and N-RNA binding in these viruses [2, 11]. RV and VSV belong to the *Rhabdoviridae* family within the *Mononegavirales* order. They share many features in both genome structure and replication strategy with the *Paramyxoviridae*. Nucleocapsid structure is somewhat different, as rhabdovirus N-RNA forms loosely coiled narrow helices approximately 24 nm in diameter, proposed to comprise 15 subunits per turn, as well as larger helices 75 nm in diameter, suggested to have approximately 53 subunits per turn [17]. *Rhabdoviridae* do not obey the rule of six and bind to nine nucleotides per N monomer rather than six. The low-resolution morphology of N-RNA is comparable to that of MeV however, comprising two globular domains. Crystallogenesis was accomplished with purified rings of recombinant N-RNA composed of ten (VSV) or eleven (RV) N monomers per ring and their structures reveal that these two N proteins adopt similar folds. Genomic RNA is encapsidated within a groove that runs along the interior of the ring, between the two globular domains of N, thus conferring protection on the RNA molecule forming a clamp of protein around it (Figure 10).

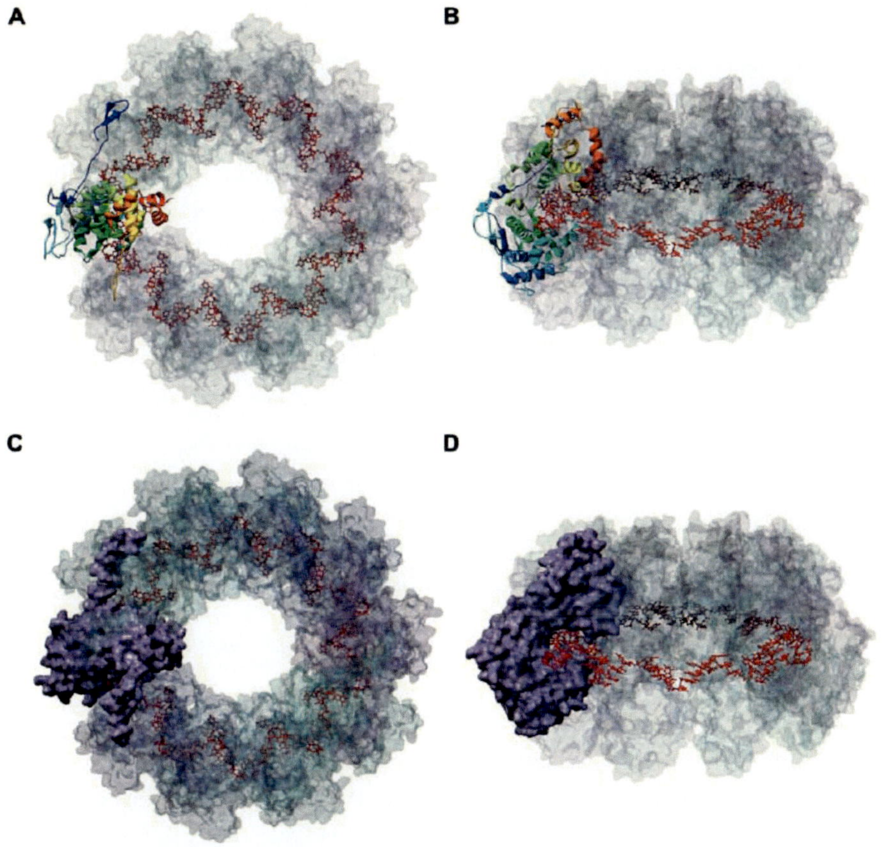

Figure 10. The structure of VSV N-RNA decameric rings. A ribbon diagram of top (A) and tilted side views (B) of VSV N-RNA shows that N comprises mostly α-helix and is divided into two distinct domains, with a groove running between them, along the interior surface of the ring that accommodates the RNA. The extreme N-terminal region forms an extended arm that interacts with the neighbouring monomer. A space filling representation emphasises the contact surfaces between adjacent N subunits (C, D).

The RNA is twisted, forming a quasi-helical structure in which, reading from the 5' end, bases one to four and base six face into the cavity at the centre of the ring, while bases five, seven and eight face into the N protein. Base six is located at the mid-point along the protomer and bulges into the central cavity. N-N interactions occur mainly between the C-terminal domains of adjacent protomers, although two additional contacts are made, one by the extreme N-terminal region that extends to form an arm around the neighbouring subunit. A flexible loop (residues 340-375 in VSV) also extends from the globular C-terminal domain to contact the neighbouring N molecule and a section of this loop is disordered in both RV and VSV. Interestingly, a serine residue (389) that is required to be phosphorylated for P binding in RV, is located in this loop, suggesting that this region may serve a similar function to N_{TAIL} in MeV, perhaps adopting secondary structure upon P-binding and influencing NC structure to expose the genomic RNA.

Ackowledgements

The author wishes to express his gratitude to all of the scientists whose work is discussed in this chapter. Particular thanks are due to Guy Schoehn (EMBL Grenoble, France) for providing figures 7 and 8, and Ed Egelman (University of Virginia, USA) for provision of software for helical reconstruction and advice. Paul Yeo (University of Durham, UK), Adam Ralph and Colin Loney (MRC Virology Unit, Glasgow, UK) are thanked for their contributions to work described and useful discussions. Finally, special thanks go to Sonia Longhi (AFMB, CNRS, Marseille) for her valuable collaboration and the kind invitation to write this chapter. Work discussed in this chapter was performed with the financial support of the European Commission, program RTD, QLK2-CT2001-01225, "Towards the design of new potent antiviral drugs: structure-function analysis of *Paramyxoviridae* polymerase" and the United Kingdom Medical Research Council.

References

[1] Adrian, M., J. Dubochet, S. D. Fuller, and J. R. Harris. 1998. Cryo-negative staining. *Micron*. 29:145-60.

[2] Albertini, A. A., A. K. Wernimont, T. Muziol, R. B. Ravelli, C. R. Clapier, G. Schoehn, W. Weissenhorn, and R. W. Ruigrok. 2006. Crystal structure of the rabies virus nucleoprotein-RNA complex. *Science* 313:360-3.

[3] Bhella, D., A. Ralph, L. B. Murphy, and R. P. Yeo. 2002. Significant differences in nucleocapsid morphology within the Paramyxoviridae. *J. Gen. Virol*. 83:1831-9.

[4] Bhella, D., A. Ralph, and R. P. Yeo. 2004. Conformational flexibility in recombinant measles virus nucleocapsids visualised by cryo-negative stain electron microscopy and real-space helical reconstruction. *J. Mol. Biol.* 340:319-31.

[5] Bourhis, J. M., K. Johansson, V. Receveur-Brechot, C. J. Oldfield, K. A. Dunker, B. Canard, and S. Longhi. 2004. The C-terminal domain of measles virus nucleoprotein belongs to the class of intrinsically disordered proteins that fold upon binding to their physiological partner. *Virus Res*. 99:157-67.

[6] Calain, P., and L. Roux. 1993. The rule of six, a basic feature for efficient replication of Sendai virus defective interfering RNA. *J. Virol.* 67:4822-30.

[7] DeRosier, D. J., and P. B. Moore. 1970. Reconstruction of three-dimensional images from electron micrographs of structures with helical symmetry. *J. Mol. Biol*. 52:355-69.

[8] Egelman, E. H. 2006. The iterative helical real space reconstruction method: Surmounting the problems posed by real polymers. *J. Struct. Biol.*

[9] Egelman, E. H., S. S. Wu, M. Amrein, A. Portner, and G. Murti. 1989. The Sendai virus nucleocapsid exists in at least four different helical states. *J. Virol.* 63:2233-43.

[10] Fooks, A. R., J. R. Stephenson, A. Warnes, A. B. Dowsett, B. K. Rima, and G. W. Wilkinson. 1993. Measles virus nucleocapsid protein expressed in insect cells assembles into nucleocapsid-like structures. *J. Gen. Virol.* 74 (Pt 7):1439-44.

[11] Green, T. J., X. Zhang, G. W. Wertz, and M. Luo. 2006. Structure of the vesicular stomatitis virus nucleoprotein-RNA complex. *Science* 313:357-60.

[12] Karlin, D., F. Ferron, B. Canard, and S. Longhi. 2003. Structural disorder and modular organization in Paramyxovirinae N and P. *J. Gen. Virol.* 84:3239-52.

[13] Longhi, S., V. Receveur-Brechot, D. Karlin, K. Johansson, H. Darbon, D. Bhella, R. Yeo, S. Finet, and B. Canard. 2003. The C-terminal domain of the measles virus nucleoprotein is intrinsically disordered and folds upon binding to the C-terminal moiety of the phosphoprotein. *J. Biol. Chem.* 278:18638-48.

[14] Mountcastle, W. E., R. W. Compans, H. Lackland, and P. W. Choppin. 1974. Proteolytic cleavage of subunits of the nucleocapsid of the paramyxovirus simian virus 5. *J. Virol.* 14:1253-61.

[15] Plumet, S., W. P. Duprex, and D. Gerlier. 2005. Dynamics of viral RNA synthesis during measles virus infection. *J. Virol.* 79:6900-8.

[16] Radecke, F., P. Spielhofer, H. Schneider, K. Kaelin, M. Huber, C. Dotsch, G. Christiansen, and M. A. Billeter. 1995. Rescue of measles viruses from cloned DNA. *Embo J.* 14:5773-84.

[17] Schoehn, G., F. Iseni, M. Mavrakis, D. Blondel, and R. W. Ruigrok. 2001. Structure of recombinant rabies virus nucleoprotein-RNA complex and identification of the phosphoprotein binding site. *J. Virol.* 75:490-8.

[18] Schoehn, G., M. Mavrakis, A. Albertini, R. Wade, A. Hoenger, and R. W. Ruigrok. 2004. The 12 A structure of trypsin-treated measles virus N-RNA. *J. Mol. Biol.* 339:301-12.

[19] Sidhu, M. S., J. Chan, K. Kaelin, P. Spielhofer, F. Radecke, H. Schneider, M. Masurekar, P. C. Dowling, M. A. Billeter, and S. A. Udem. 1995. Rescue of synthetic measles virus minireplicons: measles genomic termini direct efficient expression and propagation of a reporter gene. *Virology* 208:800-7.

[20] Spehner, D., A. Kirn, and R. Drillien. 1991. Assembly of nucleocapsidlike structures in animal cells infected with a vaccinia virus recombinant encoding the measles virus nucleoprotein. *J. Virol.* 65:6296-300.

[21] Tapparel, C., D. Maurice, and L. Roux. 1998. The activity of Sendai virus genomic and antigenomic promoters requires a second element past the leader template regions: a motif (GNNNNN)3 is essential for replication. *J. Virol.* 72:3117-28.

[22] Tarbouriech, N., J. Curran, R. W. Ruigrok, and W. P. Burmeister. 2000. Tetrameric coiled coil domain of Sendai virus phosphoprotein. *Nat. Struct. Biol.* 7:777-81.

[23] Vulliemoz, D., and L. Roux. 2002. Given the opportunity, the Sendai virus RNA-dependent RNA polymerase could as well enter its template internally. *J. Virol.* 76:7987-95.

In: Measles Virus Nucleoprotein
Editors: S. Longhi, pp. 53-98

ISBN: 978-1-60021-629-9

Chapter III

Nucleocapsid Protein Interactions with the Major Inducible 70 kDa Heat Shock Protein

*Michael Oglesbee**
Department of Veterinary Biosciences and the Department of Molecular Virology, Immunology and Medical Genetics, the Ohio State University, 1925 Coffey Road, Columbus, OH 43210, USA

Abstract

Cellular heat shock proteins (HSPs) are induced by numerous physiological stimuli including fever [98, 139]. Historical emphasis was placed upon the function of HSPs as cellular chaperones serving cytoprotective functions [52]. New roles are continually being defined, and we now know that viruses exploit HSP function to support replication in cell culture and that HSPs can significantly modulate both innate and adaptive immune responses [111]. However, the mechanisms by which HSPs enhance gene expression and replication of RNA viruses are only now being elucidated and an understanding of how HSP influences the *in vivo* outcome of virus infection is conspicuously lacking for any viral system. The present chapter will address our understanding of how the major inducible 70 kDa heat shock protein interacts with the C-terminal disordered domain on the measles virus nucleocapsid protein (N_{TAIL}) to influence both viral transcription and genomic replication. Our ability to identify structural determinants of these HSP-mediated changes in viral replication enabled the design of HSP-responsive and non-responsive measles virus variants that, in turn, allowed us to establish the biological significance of HSP-N_{TAIL} structural and functional interactions using a mouse model of measles virus encephalitis.

* E-mail: Oglesbee.1@osu.edu; Tel: 614-292-9672; FAX: 614-292-6473.

List of Abbreviations

BiP	70 kDa heat shock protein of the endoplasmic reticulum
CAT	Chloramphenicol acetyl transferase
CDV	Canine distemper virus
CD46	Membrane cofactor protein
CNS	Central nervous system
CPE	Cytopathic effect
CsCl	Cesium chloride
DnaJ	Prokaryotic homologue of hsp40
DnaK	Prokaryotic homologue of mammalian 70 kDa heat shock protein
Ed-MeV	Edmonston strain of MeV
ER	Endoplasmic reticulum
F	Viral fusion glycoprotein
GrpE	Prokaryotic nucleotide exchange factor for DnaK
H	Viral attachment (or hemagglutinin) glycoprotein
HIV	Human immunodeficiency virus
HSF-1	Heat shock factor 1
HSP70	70 kDa heat shock protein
HSPs	Heat shock proteins
hsp72	Major inducible 70 kDa heat shock protein, also known as hsp70, hsp70-1, or hspA1
hsp73	Constitutively expressed 70 kDa heat shock protein, also known as hsc70
HSV-1	Herpes simplex virus type 1
HTLV-1	Human T lymphotropic virus type 1
K_D	Equilibrium dissociation constant
L	Viral polymerase (or large) protein
MeV	Measles virus
MIBE	Measles inclusion body encephalitis
MVA	Replication defective vaccinia virus expressing T7 polymerase
N	Viral nucleoprotein
N_{CORE}	N-terminal assembly domain of N
N^0	Monomeric form of N
NR	Nucleoprotein receptor
N_{TAIL}	C-terminal domain of N
Ond-CDV	Onderstepoort strain of CDV
P	Viral phosphoprotein
PBD	Peptide binding domain
PFU	Plaque forming units
PI	Post infection
RSV	Respiratory syncytial virus
RT-PCR	Reverse transcriptase polymerase chain reaction
SeV	Sendai virus
SPR	Surface Plasmon Resonance

SSPE	Subacute sclerosing panencephalitis
SV5	Simian virus 5
TEM	Transmission electron microscopy
TTP	Tetratricopeptide
VSV	Vesicular stomatitis virus
XD	X domain of the viral P protein

1. Introduction

This section will focus on general features of 70 kDa heat shock proteins (HSPs). The discussion will show that the role of 70 kDa HSPs in support of replication for multiple virus families is a logical extension of the intracellular functions of these protein chaperones. The protein-protein interactions that underlie these functions are characterized by both high and low affinity binding reactions, and so the introduction will conclude with a discussion of the range of HSP concentrations that can be encountered within the virus-infected cell, a consideration that is paramount in establishing the range of functionally significant virus-HSP interactions that may be encountered.

1.1. Intracellular Functions of 70 kDa Heat Shock Proteins

Cellular HSPs were first recognized for their ability to facilitate protein folding [59]. By reversibly binding hydrophobic protein domains, HSPs prevent aggregation or malformation of newly formed proteins or proteins that are denatured in response to diverse cellular stressors, including transient hyperthermia or heat shock. The phrase "chaperone function" was coined because of this role of HSPs in preventing the undesirable interaction between hydrophobic domains. Multiple families of HSP provide these chaperone functions. These families differ both in molecular mass and cellular compartmentalization, with members of the 60, 70, and 90 kDa families playing a dominant role. The 70 kDa HSPs are the most abundantly expressed, representing as much as 1-2% of total intracellular proteins [74]. Expression profiles reflect cellular demands for chaperone functions, with 70 kDa HSP expression being constitutive and/or inducible. A constitutively expressed cytoplasmic isoform (hsp73, also known as hsc70) provides support of basal protein metabolism. A second isoform (hsp72, also known as hsp70, hsp70-1, or hspA1) is constitutively expressed at low levels yet is highly stress-inducible, providing similar functions as hsp73 during states of heightened protein production or protein denaturation. Expression of hsp72 is thus a salient feature of the cellular stress response, during which time hsp72 is localized to the nucleus as well as the cytoplasm. In this regard, induction of hsp72 may be viewed as a protective cellular response against diverse insult, particularly loss of cellular homeostasis that may result in (or result from) malfolding or denaturation of newly formed proteins. Although stimuli that induce hsp72 may be transient (e.g., heat shock), the elevated levels of hsp72 persist, reflecting the 48 h half-life of the protein. During this period of elevated hsp72 levels, the cell can withstand diverse protein-denaturing insults that would otherwise prove lethal. Cells protected in such a manner are considered preconditioned, and preconditioning in tissues such as brain confers protection against insults ranging from ischemia-reperfusion injury to oxygen-glucose deprivation [135, 159]. For purposes of our discussion, we will use the term HSP70 to include both hsp72 and hsp73 when addressing aspects of structure and function that are common to these isoforms. Distinctions will be made when addressing functions that may be isoform-specific.

HSP70 has a modular domain structure [68, 94]. The C-terminal binding domain of approximately 25 kDa contains a cleft exhibiting broad substrate recognition for linear sequences of 7-8 amino acids, with binding constraints mediated by the distribution of hydrophobic side groups and the nature and quantity of charged groups [43, 165]. Affinity for substrate changes dramatically in an ATP-dependent manner (Figure 1), reflecting interactions between the protein binding domain and a separate N-terminal nucleotide binding domain of approximately 45 kDa that exhibits ATPase activity [137].

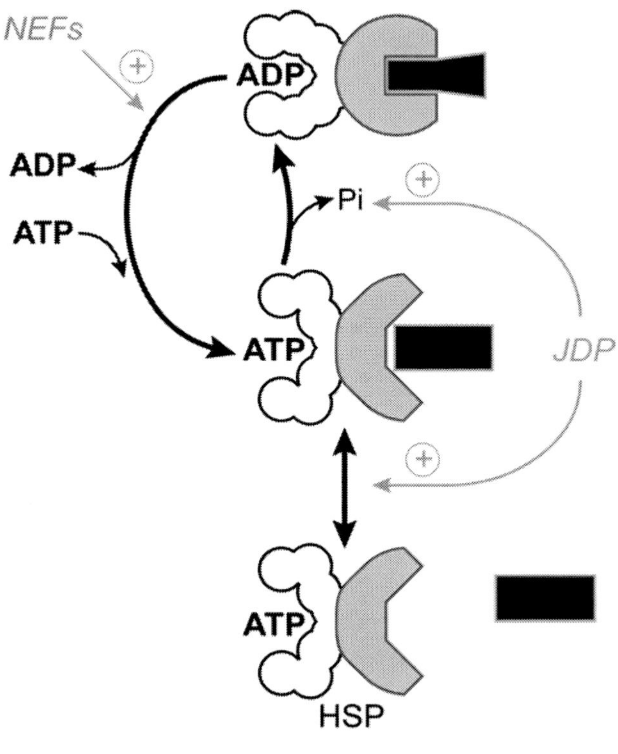

Figure 1. Modulation of HSP70-substrate interaction by nucleotides and co-chaperones. HSP70 has a nucleotide and peptide binding domain. For HSP70 bound to ATP, the peptide binding domain is open and can readily associate with (and dissociate from) substrate. ATPase activity converts ATP to ADP, resulting in conformational changes in the peptide binding domain that result in tight association with substrate. ATP-ADP nucleotide exchange assures the reversibility of this process. The client protein undergoes conformational changes as a result of HSP70 interactions, this being the driving force for folding events or functional changes in the case of HSP70-mediated activity control reactions. Co-chaperones can enhance HSP70 interactions with substrate. J-domain proteins (JDP) can promote the association of HSP70 with its substrate and stimulate HSP70 ATPase activity, locking HSP70 onto its target. Nucleotide exchange factors (NEF) facilitate replacement of ADP with ATP in the nucleotide binding domain, allowing for cycles of HSP70 binding and release that may be necessary for HSP70-dependent function.

For HSP70 bound to ATP, the peptide binding pocket is open, a conformation that promotes rapid dissociation/association with ligand [96]. Conversion of ATP to ADP causes closure of the pocket and retention of peptide, with ADP-ATP nucleotide exchange favoring repeated cycles of binding-release between the HSP and its substrate. It is this cycle of binding-release that allows HSP70 to occupy exposed hydrophobic patches that mediate undesirable protein aggregation, while allowing these same protein domains to engage in more stable intramolecular hydrophobic interactions that contribute to native protein conformation. Protein conformation is also altered as a direct consequence of HSP binding, providing the driving force required for protein maturation and renaturation. Chaperone functions of HSP70 are involved in more than just protein folding. HSP70 supports conformational changes that attend transmembrane protein transport [121], assembly and disassembly of multimeric protein complexes [6], and trafficking of binding partners into cell compartments such as the nucleus [105].

HSP70 also recognize patches of surface hydrophobicity on native proteins, resulting in altered activity of the substrate, a function known as HSP-mediated activity control [46]. In contrast to binding events involved in chaperone functions, activity control binding reactions are typically of low affinity, being primarily relevant during periods of elevated HSP70 expression within the cell. Substrate activity may either be enhanced or diminished as a consequence of HSP70 binding, reflecting changes in conformation that influence interaction with binding partners or rate of degradation. In this capacity, HSP70 modulates cell cycle progression and induction of apoptosis by binding p53 [166], kinases of the mitogen activated signal cascade [136], or apoptosis inducing factor [124]. HSP70 can also indirectly modulate cellular RNA polymerase activity by binding transcription factors or proteins that regulate function of transcription factors, such as receptor-associated protein 46, a regulator of the transactivation function of several steroid receptors [131].

1.2. Co-Chaperones and HSP70 Function

A more complete understanding of HSP70 function also requires that one consider the support of HSP70 activity provided by several classes of proteins collectively known as co-chaperones. Proteins with J domains represent a major class of co-chaperones that promote the association between HSP70 and its binding target and stimulate ATP hydrolysis by HSP70, resulting in ADP occupancy of the nucleotide binding domain and thus retention of substrate by the peptide binding domain [82]. The J domain is a conserved sequence of 70 amino acids that represents the minimal region required for association with HSP70, containing a His-Pro-Asp motif that is essential for the stimulation of HSP70 ATPase activity [65]. The remainder of the J domain-containing protein is responsible for directing HSP70 to its client protein. The J domain proteins can be subdivided into hsp40 (human Hdj1) and hsp40-like chaperones. Hsp40 is the major heat inducible J domain protein in mammalian cells and a representative member of the 40 kDa family of heat shock proteins. Hsp40-like proteins are either structurally homologous to hsp40 or the corresponding prokaryotic DnaJ (i.e., type I hsp40-like proteins) or structurally distinct, representing proteins that appear to have recruited the J domain in order to fulfill a number of specific functions (i.e., type II hsp40-like proteins) [65]. The HSP70-hsp40 partnership is invoked in both protein folding and refolding events. Substrate recognition by hsp40 is broad, where surface hydrophobicity of a large number of partially folded (or unfolded) proteins is recognized by a peptide binding cleft that is functionally analogous to that of HSP70 [86]. Hsp40-like proteins tend to exhibit a more restrictive specificity for HSP70 substrates. For example, auxillin is an hsp40-like protein of cellular origin that directs hsp73 to clathrin; thereby supporting cycles of clathrin assembly and disassembly associated with receptor mediated endocytosis [156]. Polyomavirus T antigens are hsp40-like proteins that regulate viral DNA replication, transcription, and tumorigenesis. The T antigen may direct hsp73 to members of the retinoblastoma family of cellular proteins in order to mediate the release and thus activation of the E2F family of transcription factors [47]. These transcription factors promote viral replication and tumorigenesis by inducing expression of cellular genes required for entry into the S phase of the cell cycle [21].

Replacement of ADP with ATP in the nucleotide binding site of HSP70 is needed to facilitate release of the substrate, thereby supporting the cycles of HSP70 association/dissociation with substrate that is characteristic of HSP70-mediated folding reactions. The importance of this function is underscored by the fact that nucleotide exchange is rate limiting for substrate release by HSP70 [147]. The exchange necessarily enhances the ATPase activity associated with HSP70 and is thus considered an integral part of the HSP70 ATPase cycle [134]. Nucleotide exchange factors that enhance chaperone function of HSP70 include members of the 110 kDa heat shock protein family [37] and the family of Bag proteins [17, 141]. The Bag family consists of at least six members sharing a common C-terminus that directly interacts with the nucleotide binding domain of HSP70 [143]. Although not essential for HSP70's role in folding reactions, Bag proteins have been identified as either positive or negative regulators of HSP70 function depending upon concentration [48]. In general, we may view Bag proteins as positive regulators since Bag concentrations are low under physiological conditions (i.e., in substoichiometric amounts relative to HSP70) and at these low concentrations and in the presence of inorganic phosphate will enhance HSP70 function [48]. Like J domain proteins, Bag interacts with numerous other cellular binding partners that influence diverse cellular processes that include apoptosis and cellular differentiation, providing a putative link through which HSP70 may influence these cellular events [142]. Collectively, modulators of nucleotide exchange may be viewed as a compliment to the J domain proteins. Stimulation of ATPase activity by J domain proteins would promote ADP occupancy of the nucleotide binding site, with ADP-ATP nucleotide exchange facilitated by hsp110 and Bag proteins providing the balance that is needed in order to sustain the HSP70 ATPase cycle and thus the reversible interaction of HSP70 with client proteins.

Additional modulatory proteins have been identified that operate outside of the HSP70 ATPase cycle. The 43 kDa Hip competitively inhibits the binding of Bag-1 to the nucleotide binding domain of hsp73 [71]. Hip, together with the co-chaperones Hop and Chop, also contain a tetratricopeptide (TTP) motif that mediates interactions with the C-terminal EEVD motif found on HSP70s as well as hsp90 [130]. This allows the TTP-containing co-chaperones to facilitate the cooperative interaction between chaperone complexes. For example, Hop coordinates the interaction between HSP70 and hsp90 in driving conformational regulation of steroid receptor signal transduction [66]. Here, the co-chaperone may mediate complex assembly as well as facilitating the exchange of a client protein from one chaperone to another within the complex. Inhibition of HSP70 function by Chip may allow a substrate to be released from a folding pathway and thereby enter into a degradation pathway [31], or release heat shock transcription factors from inactive complexes with HSP70 to result in transcription of heat shock protein genes [34]. The existence of co-chaperones does not imply that all roles of HSP70 require or are regulated by such molecules, only that they have to be considered when evaluating the functional requirements of a specific HSP70 interaction with a target molecule.

In general, the degree to which co-chaperones are involved in supporting HSP70-virus interactions have not been adequately addressed for most viral systems [92]. It is possible that these cofactors play major roles in the replication of some viruses (particularly where chaperone functions of HSP70 are employed) and yet no significant role in the HSP70-dependent aspects of replication for other viral families. The difference may relate to the degree to which the virus has evolved to exploit a host cell protective function (i.e., the expression of HSP70). Alternatively, it may be that levels of co-chaperones have not been rate limiting in functional analyses focused upon HSP70-virus interaction and thus their participation has gone undetected. Co-chaperones do have the potential to be rate limiting to virus-HSP interaction, this being suggested by the observation that overexpression of Bag-1 enhances the transcriptional activity of human polyomavirus promoters [140]. However, the majority of *in situ* assays or cell-free assays utilizing soluble cell extract are likely to contain sufficient levels of co-chaperones to support HSP70 function over a range of concentrations. Moreover, co-factors may be rendered non-essential if the level of HSP supplementation is sufficiently high [2].

1.3. Roles of 70 kDa HSPs in Viral Replication

The multiple roles of HSP70 in protein metabolism and the abundance within cells underscores their potential significance in viral replication, this being the focus of a recent review [92]. For every step in the viral replication cycle (i.e., attachment/penetration, uncoating, transcription and genome replication, and virion morphogenesis), one can cite an example of a viral system that draws upon HSP70 for support. Hsp70 can be expressed on the cell surface where it may support viral entry. Hsp70 is not expressed as an integral membrane protein, but rather attached to a number of different cell surface proteins that may include Toll-like receptors 2 and 4, CD36 and SR-A scavenger receptors, low-density lipoprotein receptor related protein (CD91), C-type lectin receptor LOX-1, and the co-stimulatory molecule CD40. See Asea [4] for a review of cell surface molecules that may be utilized by HSP70 for attachment. A mechanism for export of HSP70 has not been established, although the HSP could readily be acquired from the extracellular milieu, being released by both stressed and necrotic cells. Enveloped viruses utilize membrane glycoproteins to mediate fusion between the viral envelop and plasmalemma or between the plasmalemma of an infected and uninfected cell in order to assure the transmission of infection. Cell surface hsp73 may be essential to this fusion function as in the case of human T lymphotrophic virus type 1 (HTLV-1) gp46 [127]. In the case of non-enveloped viruses such as rotavirus and adenoviruses, hsp70 is required for uncoating and/or transport of the viral genome to sites of transcription and replication [118, 129].

HSP70 is directly involved in transcription and/or replication of viral genomes. Mechanistically, this aspect of virus-HSP70 interaction is best understood for DNA viruses and was first characterized in bacteriophages. Lambda phage recruits the E.coli homologue of HSP70 (i.e., DnaK) to stimulate transcription and genome replication by removing inhibitory viral proteins from replication complexes [2]. Prokaryotic homologues of Hsp40 (i.e., DnaJ) and a nucleotide exchange factor (GrpE) are an integral part of this process at low ratios of DnaK to client protein, whereas DnaK can function in a GrpE-independent manner when DnaK is present in excess of its substrate [2]. In contrast, hsp72 promotes efficient viral DNA transcription and replication of human papillomavirus by promoting assembly of preinitiation complexes on the origin of DNA replication; these complexes contain the viral E1 helicase, and E1 binding to the origin is weak in the absence of hsp72 [88]. Analogous interactions occur between large tumor antigens of polyomaviruses and hsp73 [20], or the herpes simplex virus type 1 (HSV-1) initiator protein UL9 and hsp72 [144], with assistance from J domain co-chaperones incriminated in the latter case as well. These observations may provide a mechanistic basis for the *in vitro* observation that heat shock stimulates reactivation of latent HSV-1 [57]. The ability of heat shock or selective hsp72 overexpression to enhance adenovirus replication and cytopathic effects (CPE) in cell culture has been shown, although the underlying mechanisms are unknown [61]. Retroviruses may properly be considered together with DNA viruses in this discussion, since an explanation for the heat shock-induced stimulation of retroviral gene expression [3, 50, 60] may be found by analogy to hepadnaviruses; hsp73, the co-chaperone Hdj1, and hsp90 combine to directly activate the hepatitis B virus reverse transcriptase to promote initiation of viral DNA synthesis from pregenomic RNA [7]. HSP70 can also have indirect effects upon viral gene expression. This would include HSP70-dependent effects upon cellular transformation that enables gene expression by some DNA viruses, the support of cellular kinase activities that may be required for viral transcription, and stabilization of the virus-infected cell thereby prolonging the interval during which the cell can actively support viral gene expression. This cell preservation phenomenon could reflect the role of HSP70 in maintaining cellular homeostasis and in mediating the inhibition of apoptosis.

Chaperone functions of 70 kDa HSPs can be subverted in the basal support of viral replication by promoting maturation of viral membrane glycoproteins and core proteins, assembly of nucleocapsid, and trafficking of viral subunits between cell compartments. Associations between BiP, a 70 kDa HSP family member restricted to the endoplasmic reticulum (ER), and early folding intermediates of viral membrane glycoproteins demonstrate that 70 kDa ER chaperones are part of the normal viral glycoprotein maturation pathway. BiP binds the polypeptide backbone of the measles virus (MeV) fusion and hemagglutinin glycoproteins whereas the ER chaperones calnexin and calreticulin interact with the carbohydrate moieties [11]. Such transient interactions have been documented in the maturation of glycoproteins for several viruses including human immunodeficiency virus (HIV) [113], rabies virus [49], vesicular stomatitis virus (VSV) [58], influenza virus [119], and Sendai virus (SeV) [126]. Cytosolic HSP70 can also be involved in membrane glycoprotein maturation as in the case of hepatitis B virus. Here, HSP70 contributes to a mixed topology where the N-terminus of the L protein can either be cytosolic and serve as a matrix protein or extracellular and serve as an attachment protein [81]. Such a role is more the exception than the rule and does not overshadow the role of BiP in L glycoprotein maturation [28].

The fidelity of nucleocapsid assembly by the non-enveloped polyomavirus is mediated by hsp73 interactions with the C-terminal disordered domains of VP1 [29, 32, 95]. Similarly, the reovirus sigma one core protein that is responsible for attachment is also dependent upon HSP70 to assure appropriate maturation into the trimeric form [85]. For other systems, a link between HSP70 and nucleocapsid assembly is more speculative, based upon HSP binding to or colocalization with capsid proteins. Such is the link between hsp72 and newly synthesized capsid precursor P1 of poliovirus and coxsackievirus B1 [91], the adenovirus type 5 fiber protein [90], or the nucleocapsid protein (N) of the morbilliviruses MeV and canine distemper virus (CDV) [105, 163]. For MeV and CDV, hsp72 also mediates the intranuclear translocation of the viral N protein to form eosinophilic inclusion bodies in that location [105]. The ultrastructural basis for these structures in CDV infected cells are nucleolar derivatives known as nuclear bodies, and N protein content may be sufficient to support formation of nucleocapsid helices [104]. It is unknown in this instance if nuclear body formation has any influence upon viral replication. The impact of HSP70 on viral protein trafficking in direct support of replication has been shown for HIV-1, where hsp72 mediates translocation of pre-integration complexes from the cytoplasm to the nucleus [1].

For all of these viral systems, the *in vivo* significance of HSP70-mediated alterations of viral gene expression remains to be shown. A direct role for HSP70 in gene expression for RNA viruses lacking a proviral DNA intermediate is also conspicuously lacking, with the exception of CDV and MeV. For these viruses, hsp72 but not hsp73 enhances both transcription and genome replication, with at least part of this effect mediated by the binding of hsp72 to the C-terminal disordered domain of the N protein (N_{TAIL}). The following sections will focus on how hsp72/N_{TAIL} interactions can enhance *Morbillivirus* gene expression in the MeV system. Finally, I will describe how the understanding of the molecular basis of the hsp72/N_{TAIL} interaction led us to design hsp72 responsive and non-responsive MeV variants that can then be used to dissect the biological (i.e., *in vivo*) significance of MeV-hsp72 interactions in a mouse model of MeV encephalitis.

1.4. Levels of HSP72 in the Virus-Infected Cell

Any functional analysis of virus-hsp72 interaction must take into account the basal levels of hsp72 that are present within the system, then increase or suppress these levels with the aim of detecting hsp72-dependent changes in a viral replication parameter. Ablation of hsp72 expression is problematic in terms of cell survival, requiring use of inducible suppression in stabilized (i.e., non-stressed) cell populations. Supplemental increases in hsp72 must take into account the variable basal expression that can be exhibited between different culture systems or between different animal species, this defining the relative increase above background. High level constitutive expression of hsp73 is virtually universal, whereas hsp72 is primarily stress inducible. Primate cell lines and tissues are characterized by significant constitutive expression of hsp72 whereas basal levels are lacking in rodent cell lines and tissues. Even with heat shock induction of hsp72 in rodent brain, the levels are only one tenth of that detected in non-diseased brains of humans [115]. Constitutive and stress inducible hsp72 levels also tend to be increased in primary culture relative to the tissue giving origin to those cultures, greater in continuous cell lines relative to primary cultures of the same cell lineage, and greater in undifferentiated versus differentiated cells of the same lineage [100]. Viral dependence upon hsp72 in support of replication could explain why continuous cell lines of primate origin (e.g., HeLa, 293, Vero) are so highly permissive to tissue culture-adapted MeV infection, in contrast to the more limited permissiveness of rodent cell lines. Incrimination of hsp72 in support of viral replication can rely upon heat shock to induce hsp72, but that induction should be documented as there are cell growth conditions that determine the degree to which further increases in hsp72 levels may occur. Growth of cells under stressful (i.e., non-homeostatic) conditions will elevate basal hsp72 levels, rendering them stress preconditioned and thus refractory to further elevations in hsp72 [146].

Virus infection is a potent inducer of hsp72. This has been demonstrated as a general phenomenon for both RNA and DNA viruses, and is generally accompanied by the induction of endoplasmic reticulum 70 kDa HSP (i.e., BiP) [11, 30, 45, 105, 120, 153, 157, 160]. These findings have been primarily based upon analysis of virus infected cells *in vitro*, including our work with MeV-infected Vero and murine neuroblastoma cells [111]. Evidence of hsp72 induction following *in vivo* infection is restricted to our characterization of CDV-infected dog brain using dual label confocal immunofluorescence microscopy [105]. In that report, expression of nucleocapsid protein was used as a marker for viral infection. Hsp72 expression was increased in virus infected cells relative to uninfected cells within the same brain, supporting a direct mechanism of virus induction.

One mechanism for direct viral induction of hsp72 would be production of a large number of viral proteins during the course of virus replication, thereby inducing an unfolded protein response by the cell. In the cytosol, the accumulation of insoluble viral proteins consumes available stores of hsp72 that are otherwise engaged in binding to heat shock factor 1 (HSF-1), a transcription factor mediating the induced expression of hsp72. Redistribution of hsp72 to viral targets liberates HSF-1 to stimulate production of hsp72 transcripts [5]. Herpesvirus-induced hsp72 is thought to represent such a mechanism [75, 112]. Respiratory syncytial virus (RSV) infection has been shown to cause redistribution of HSP70 to cytosolic lipid rafts in association with viral nucleocapsid particles, although HSP induction was not demonstrated in the Hep2 cells that were the basis for that work [19]. The experimental approach did not permit distinction between hsp72 and hsp73. Failure to show hsp72 induction illustrates the significance of culture systems used to study virus-HSP interactions since RSV causes both hsp72 redistribution and induction in A549 (alveolar type II-like epithelial) cells, a cell system that more closely models the natural target of RSV [16]. Similar unfolded protein responses occur in the endoplasmic reticulum (ER), where expression of viral membrane glycoproteins induces BiP/grp78. The induction of HSPs in this location is essential to maintain homeostasis and if that cannot be done, apoptosis will ensue [158]. The stress of a large number of nascent viral glycoproteins that require chaperones is detected by resident transmembrane proteins activating one of three pathways, each pathway inducing different subsets of ER chaperones. These sensors include the activating transcription factor 6 (ATF6), inositol-requiring enzyme 1 (IRE1), and PKR-like ER kinase (PERK) [132]. Accumulation of the SARS coronavirus spike protein induces activation of the PKR-like ER kinase [26]. *Paramyxoviridae* members, including MeV [11], simian virus 5 (SV5), and SeV [117] induce an ER unfolded protein response, although the signaling pathways are not characterized.

Viral infection can also induce HSP70 very early during the course of viral infection when the stimulus for an unfolded protein response is lacking. Binding of HIV gp120 to cell surface receptors results in the activation of signal transduction pathways that may or may not be mediated by HSF-1 [45, 153]. Selective induction of hsp73 is mediated by SV40 large T antigen, where the transactivating viral protein recognizes hsp73-specific promoters by interacting with promoter-specific transcription factors [35]. Our *in vitro* work with MeV indicates that hsp72 is induced early in the course of infection and cannot be attributed to an unfolded protein response [111]. Western blot analysis of Edmonston MeV infected murine neuroblastoma cells showed hsp72 in virus infected but not in uninfected cell total protein. This hsp72 was induced approximately 12 h prior to detection of viral N protein in cell extracts. Moreover, hsp72 levels peaked at 24-36 h post-infection (PI) and then began to decline, whereas N protein levels continued to rise through 60 h PI. Immunocytochemical staining of viral N protein and hsp72 showed that cellular contact with UV-inactivated virus is sufficient to induce hsp72, suggesting that virus-receptor interactions trigger an HSP induction pathway that would be similar to that described for HIV (Oglesbee et al., unpublished observation). The kinetics of induction would make hsp72 available for incoming nucleocapsid in support of primary transcription as well as transcription mediated by newly formed templates and polymerases. Both MeV and CDV cytosolic nucleocapsid protein aggregates are extensively colocalized with hsp72.

Is this virus induction necessary to support virus replication? Our laboratory has used siRNAs to inhibit MeV-induced hsp72, and this treatment results in diminished support of MeV gene expression (Oglesbee et al., unpublished observations). In that work, reduced MeV permissiveness was also observed in HSF-1 knockout mouse embryo fibroblasts, where both virus and heat-shock induction of hsp72 is blocked. The reliance on virus-induced HSP70 levels or redistribution is also exhibited by the avian adenovirus CELO. CELO encodes a protein (Gam 1) that leads to nuclear accumulation of both HSP70 and the co-chaperone Hdj1. This translocation is essential for viral replication in that deletion of Gam 1 is lethal to the virus unless cells are heat shocked, a treatment causing cytosolic to nuclear redistribution of HSP70, or transfected to express Hdj1 so that the J domain protein can direct HSP70 to the nucleus [53]. Other evidence for the necessity of virus induced changes in HSP70 levels or distribution are more circumstantial. Cell line differences in hsp72 inducibility have been correlated to differences in permissiveness to CDV infection [106]. Two Vero cell lines were identified that differed in the ability to support production of high titer viral progeny. The more permissive line was characterized by a greater degree of heat shock inducibility of hsp72 and also exhibited a greater proportion of cells in the S phase of the cell cycle during log phase growth. Hsp72 inducibility is maximal during the S phase of the cell cycle [93]. Permissiveness to infection also declined with high cell passage and this decline was also correlated to diminished hsp72 inducibility and S phase cell cycle compartmentalization. It is likely that similar variability in basal and inducible hsp72 levels would influence patterns of virus replication *in vivo* as a function of tissue compartment. Others have noted hsp72 in the virion, using this to support the importance of hsp72 early during the course of infection [56]. We have detected hsp72 in SeV particles, but similar analyses have not been performed for MeV (Oglesbee et al., unpublished data).

Indirect mechanisms by which virus infections would increase hsp72 levels would be operative *in vivo*. Febrile responses are a characteristic feature of MeV infection, and temperatures as low as 39°C have been shown to be potent inducers of hsp72 in human brain [98]. Increased plasma levels of inflammatory cytokines are also a characteristic of infection, and both IL-1 [138] and TNF-α [63] induce hsp72. The result would be elevated hsp72 expression in all organ systems and within uninfected as well as infected cells, increasing the levels that the virus encounters as infection spreads. A similar scenario is created *in vitro* when cells are heat shocked 12 h prior to infection.

There is a difference in levels of hsp72 that are available to the virus when comparing viral versus heat shock induction. Biochemical analysis of purified cytosolic nucleocapsid of CDV and MeV indicates that more hsp72 is available for nucleocapsid interaction when cells have been transiently shocked 12 h prior to infection (i.e., preconditioned). Complex formation between hsp72 and nucleocapsid can be demonstrated in virus-infected non-shocked cells, but the degree of complex formation is much greater in preconditioned cells [106, 162]. The reason for the difference likely reflects differences in the distribution of hsp72 between soluble versus insoluble cell fractions that, in turn, reflect the degree to which hsp72 is engaged in high affinity target binding. Transient thermal stress induces an unfolded protein response, and the induced chaperones return the cell to homeostasis after the proper folding of nascent cellular proteins is rescued or terminally denatured proteins are targeted for degradation. This process is likely complete within 6h post-shock since this is the interval after which induction of hsp72 is generally terminated. Beyond this point, the induced hsp72 remains for a considerable period of time, reflecting the 48 h half life of the protein [97]. Without a binding partner, this hsp72 is available for interaction with viral targets, being abundant in the soluble cell fraction [110]. A similar scenario would be created when constitutive expression of hsp72 is driven by a plasmid vector. Accordingly, both heat shock-induced and enforced constitutive expression of hsp72 results in identical structural and functional interactions with CDV and MeV.

In summary, hsp72 is a fundamental component of the environment in which viral replication occurs. Viral induction of hsp72 assures this in cases where constitutive cellular expression may be lacking. Only cell-free assays or strategies that selective suppress hsp72 can be expected to be hsp72-depleted. These low levels of hsp72 are contrasted to the higher available levels that would be encountered when a stress preconditioned cell is infected. Our subsequent discussions must take into account the availability of hsp72 when considering viral binding targets that differ in hsp72 affinity, and the functional consequences of those interactions.

2. HSP72 and *Morbillivirus* Gene Expression

2.1. Binding of HSP72 to *Morbillivirus* Nucleocapsid

The initial observation that hsp72 interacts with the *Morbillivirus* nucleocapsid was made with the Onderstepoort strain of CDV [106, 107]. Nucleocapsid was purified from the cytoplasm of infected Vero cells. The final CsCl isopycnic gradient step yielded particles of high purity, although two nucleocapsid density variants were recovered from highly permissive cells, referred to as dense (1.30 gm/ml) and light (1.29 mg/ml) nucleocapsid (Figure 2). Biochemically, the two appeared identical except for the additional 70 kDa host protein that was consistently co-purified with light nucleocapsid. The protein was recovered in variable amounts, representing 4-8% of total protein in the nucleocapsid preparation. A laboratory accident, i.e. an incubator malfunction that resulted in a transient heat shock of the infected cells, led to the identification of the 70 kDa protein as hsp72. With the heat shock, only light nucleocapsid was recovered and the content of 70 kDa host protein was dramatically increased. Subsequent experiments reproduced the phenomenon using a controlled heat shock that was performed prior to infection in order to separate the induction stimulus from the elevation in cellular levels of hsp72. The hsp72 was in molar excess of the L protein which is normally 2% of the CDV nucleocapsid total protein, and frequently in excess of the amount of P protein which is normally 8% of the nucleocapsid total protein. This made the nucleocapsid (N) protein the probable binding target of hsp72.

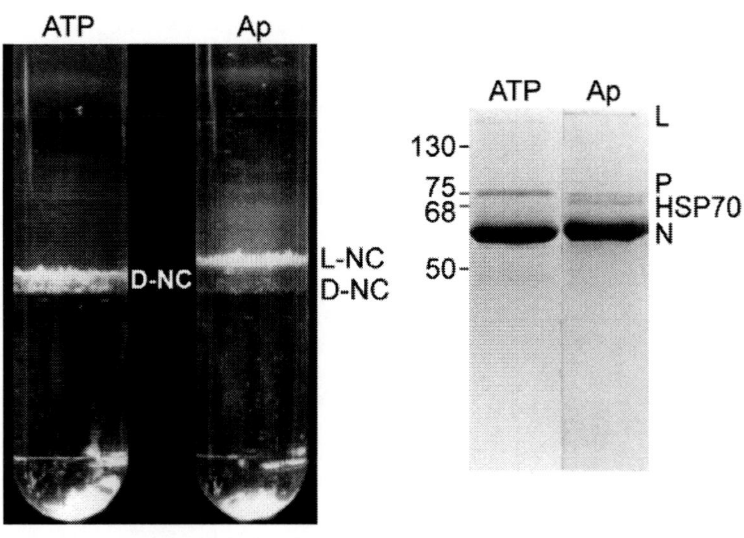

Figure 2. ATP-dependent reversible interactions between hsp72 and nucleocapsid of Onderstepoort CDV. Nucleocapsid is purified from the cytoplasm of infected Vero cells using CsCl isopycnic density gradients (left). In the presence of supplemental ATP, only dense nucleocapsid (D-NC) is recovered. Analysis of total protein by SDS-PAGE following by Coomassie brilliant blue staining (right) illustrates the expected proportions of the viral N, P, and L proteins. When ATP is depleted in the cytoplasmic extracts using apyrase treatment (Ap), light nucleocapsid (L-NC) is recovered as well as dense, and total protein analysis reveals the association with HSP70.

Identification of the N protein as a putative binding target was also consistent with the extensive colocalization of hsp72 and N protein but not P protein within CDV infected cells both in vitro and *in vivo* [105]. Nucleocapsid of the Hallé strain of MeV was subsequently shown to behave in an identical manner, supporting the more general applicability of these findings [150].

The association of hsp72 with the CDV and MeV nucleocapsid is ATP-dependent and reversible, consistent with functional hsp72-substrate interactions (Figure 2) [110, 150]. The original nucleocapsid isolation experiments did not control for ATP levels. Advancing cytopathic effect (CPE) can result in a decline in ATP concentration within the cytosol, although measurements of ATP in cytosolic extracts at times typical for nucleocapsid harvest show no significant difference in concentration between infected and uninfected control cells. The hsp72 nucleotide binding site could thus be occupied by either ATP or ADP, and it would only be the hsp72 subset containing ADP that would be bound to nucleocapsid with an affinity sufficient to withstand the rigors of a CsCl gradient. The degree of hsp72-nucleocapsid interaction that is detected in purified samples is therefore only a fraction of that occurring in situ. Apyrase treatment of cytoplasmic extracts significantly enhances the degree to which hsp72-nucleocapsid complexes can be recovered regardless of hsp72 levels within the infected cell. Apyrase dephosphorylates ATP, forming ADP and AMP, and eliminates detectable ATP in cytoplasmic extracts from starting levels of 5-10 μM. Once ATP is depleted (and the levels of ADP enhanced), hsp72 as well as hsp73 can be identified not only in the light nucleocapsid variant, but also within the dense nucleocapsid variant. Conversely, ATP supplementation to 2.5 mM eliminates recovery of hsp72-nucleocapsid complexes, consistent with the low substrate binding affinity of hsp72 when ATP occupies the nucleotide binding site. This treatment also eliminates recovery of light nucleocapsid, indicating that expression of this density variant is hsp72-dependent.

2.2. Functional Interactions between HSP72 and *Morbillivirus* Nucleocapsid

Functional significance of hsp72-nucleocapsid interaction has been shown for MeV and CDV in both virus-infected cells and in cell-free assays. Within cells, changes in the infection phenotype are mediated by transient heat shock or selective hsp72 overexpression following either transient or stable transfection with an expression construct. Effects are identical regardless of whether the increased hsp72 levels are part of a cellular stress response or selective hsp72 over-expression. Either treatment increases cellular levels of viral transcripts and genomic RNA (Figure 3) [24, 109, 151, 162]. This can be shown following lytic infection of preconditioned cells or induction of the cellular stress response in cell lines supporting stable persistent infection. Increased viral transcript levels result in increased viral protein expression, including that of the viral F and H glycoproteins that are responsible for much of the virus-induced CPE, particularly syncytia formation [64, 109, 150, 151]. The level of F and H expression is proportional to plaque area formed on infected cell monolayers, and hsp72 supports the emergence of large plaques where small plaque areas are otherwise formed. Induction of hsp72 also causes the formation of syncytia in stable non-cytopathic persistent infections (Figure 4).

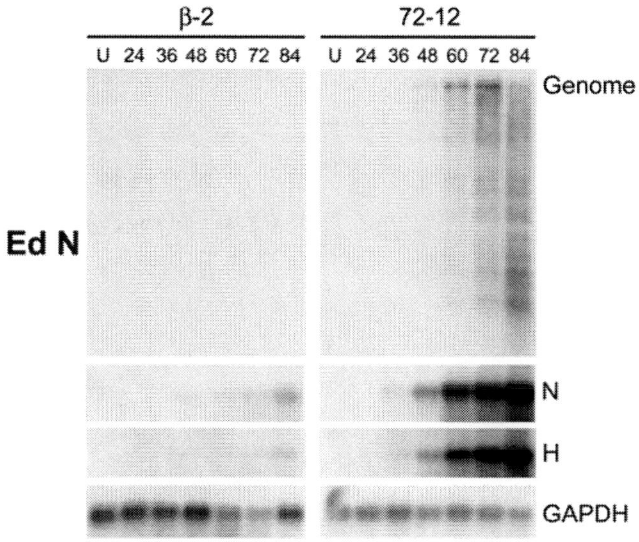

Figure 3. Selective overexpression of hsp72 in stably transfected murine neuroblastoma cells (72-12) supports increased Edmonston MeV genome replication and transcription. Northern blot analyses contrast the RNA expression levels to those observed in a vector transfected control cell line (β-2) over a time course that spans 24-84 h PI. Analysis of cellular GAPDH transcripts was used as a loading control and the negative control was total RNA from uninfected neuroblastoma cells (U) [24].

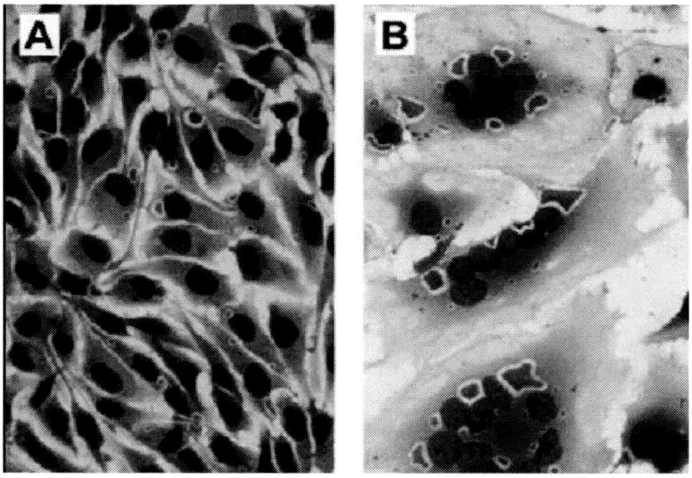

Figure 4. Heat shock induction of hsp72 converts a stable persistent infection to a lytic state. Mink lung cells (CCL64) support a stable persistent infection by a raccoon isolate of CDV (RCDV), where CPE is restricted to formation of discrete cytoplasmic inclusion bodies. Within 24 h post heat treatment, CPE is characterized by extensive syncytia formation that was correlated to increased hsp72-nucleocapsid complex formation and cell-free transcriptional activity, increased transcript accumulation within infected cells, and increased viral protein expression, particularly that of the fusion glycoprotein [109].

Increased hsp72 levels beyond that otherwise induced by the virus result in increased progeny release, but the change is greatest in cell lines where basal hsp72 is lacking (e.g., murine neuroblastoma cells) [24, 106, 109, 150, 151]. The effect is more difficult to demonstrate in Vero or HeLa cells, suggesting that viral protein expression and genome replication is not a significant rate limiting factor to progeny release in these cell lines. Primary cultures have not been examined for an effect of hsp72 on progeny release.

Hsp72-dependent stimulation of viral transcription and genome replication can be attributed to increased nucleocapsid polymerase activity, and this has been demonstrated for both CDV and MeV using cell-free nucleocapsid transcriptional assays [109, 110, 150, 151]. These assays measure elongation of pre-initiated transcripts and therefore do not gage genome replication, a process that would require co-expression of viral N protein to encapsidate the nascent genome. Synthesis of genomic/antigenomic RNA and encapsidation of those growing strands are integrally linked. The cell-free reactions utilize soluble cell extracts that would be a source of additional HSPs that could engage in reversible nucleocapsid binding reactions as well as a source of any co-chaperones that may be required. Nucleocapsid can either be CsCl gradient purified or contained within infected insoluble cell fractions. Purified nucleocapsid is capable of only limited enzymatic activity that can only be measured by the ability to incorporate nucleotide into precipitable nucleic acid. Activity is higher when insoluble cell extracts are used as a source of nucleocapsid, allowing the gene identity of newly formed transcripts to be defined by electrophoretic mobility and hybridization to viral gene specific probes. Using this approach for CDV, we showed that purified light and dense hsp72-associated nucleocapsid exhibited polymerase activity, while nucleocapsid lacking hsp72 do not [110]. Activity is significantly greater in the hsp72-containing dense relative to light nucleocapsid. Similar results are obtained when polymerase activity is measured in insoluble cell fractions from CDV and MeV-infected cells. Transcript production is increased in insoluble fractions derived from cell expressing elevated hsp72 levels and antibody against hsp72 inhibits viral transcriptional activity whereas antibody against hsp73 has no effect, supporting the isoform-specificity of this 70 kDa HSP function [110]. A similar approach was recently used to document a role for HSP70 in RSV polymerase activity, using cytosolic lipid rafts from infected cells as a source of nucleocapsid [19].

Addition of purified hsp72 or mixtures of hsp72 and hsp73 to infected cell extracts result in a dosage-dependent stimulation of cell-free transcription up to a final reaction concentration of 2.0 μg/ml HSP70 [110]. This level of supplementation is the estimated concentration of non-peptide bound HSP70 in the soluble fraction of Vero cells following induction of the cellular stress response. Transcription is inhibited above this concentration. Similar biphasic dose responses have been demonstrated when 70 and 90 kDa HSPs are used to rescue enzyme activity of thermally denatured luciferase [133]. Inhibition of enzyme function at supraphysiological levels of heat shock proteins thus appears to be more general phenomenon that is in contrast to support of those functions at physiological levels. Supplementation of purified hsp73 is without effect on CDV. Addition of purified hsp72 to purified nucleocapsid stimulates activity approximately 6-fold above background, although the level of activity never achieves that observed in pre-formed hsp72-nucleocapsid complexes (i.e., 22-fold above background) and does not exhibit the biphasic dose response observed with insoluble cell extracts. These findings suggest that purified nucleocapsid lacks the platform required for optimal polymerase function. In the case of RSV, that platform would be a lipid raft. In the case of MeV or CDV, that platform may be a cytoskeletal scaffold. We cannot rule out the possibility that insoluble fractions contribute co-factors for viral polymerase function or co-chaperones required for hsp72-mediated effects. It also remains to be shown the degree to which these findings, which are based upon viral transcriptase activity, applies to the replicase function that is the basis for genome synthesis.

2.3. Functional Implications of HSP72-Associated Nucleocapsid Conformations

The hsp72-mediated effect upon nucleocapsid buoyancy is accompanied by ultrastructural changes that suggest a basis for hsp72-dependent stimulation of transcription and genome replication [107, 150]. The light nucleocapsid variant of CDV exhibits extensive nuclease susceptibility of the genome that is accompanied by ultrastructural evidence of helical expansion [107]. In the absence of hsp72, the RNA genome within nucleocapsid is inaccessible as demonstrated by its resistance to ribonuclease degradation. The inaccessibility should represent a hindrance to viral polymerase function since the RNA template remains encapsidated during both transcription and genome replication. Thus, hsp72 may stimulate transcription and genome replication by altering nucleocapsid conformation in a way that enhances exposure of the genome.

Intermediate resolution studies of the MeV nucleocapsid provide insight as to how hsp72 may alter nucleocapsid conformation. Bhella et al. used negative stain transmission electron microscopy (TEM) to show that Ed-MeV produces nucleocapsid helices of multiple conformations with varying degrees of flexibility [8, 9] (see also Chapter 2). Recombinant N protein was expressed from baculovirus vectors, where nucleocapsid-like particles form spontaneously around cellular RNAs. The resultant structures are morphologically indistinguishable from nucleocapsid formed by infectious virus and exhibit similar densities on CsCl gradients. As with nucleocapsid isolated from the cytoplasm of CDV infected mammalian cells, subnucleocapsid rings are shed from the parent structures, a fortuitous event in that it facilitates calculation of the number of N protein subunits per turn based upon projection averages. Cryo-negative stain TEM is an advanced technique that better preserves native three-dimensional structure, in contrast to the flattening of particles that can occur during drying for conventional TEM. Bhella et al. used this technique to further evaluate MeV nucleocapsid morphological variation. Image processing is used to sort nucleocapsid according to diameter, helical pitch, and number of subunits per turn, generating three-dimensional reconstructions for each variant. Cross-sections of the longitudinal reconstructions suggest a superficial location of the RNA binding groove on the amino-terminal portion of N that is responsible for driving helix formation (i.e., N_{CORE}). This superficial location should permit access of the polymerase to genomic RNA with minimal perturbation of helical structure. Accessibility of the RNA would be regulated by the intrinsically disordered C-terminus of N (N_{TAIL}) whose surface projections would mask the genomic RNA. N_{TAIL} conformation would determine the degree to which the underlying genome is masked (or unmasked). In addition to this direct influence of N_{TAIL} on RNA accessibility, we must consider indirect effects where an N_{TAIL} conformation may induce changes in the helical pitch of N_{CORE} projections that could enhance exposure of the RNA binding groove. N_{TAIL} cannot be visualized by these imaging techniques due to its inherent structural disorder, complicating the identification of the mechanisms that may be involved.

The intrinsic disorder of N_{TAIL} allows it to interact with a variety of partners and adopt multiple conformations, each representing a potential opportunity to expose genomic RNA [13, 15, 89]. At least one of the N_{TAIL} conformations destabilizes the interaction between turns of the nucleocapsid helix, since selective proteolytic removal of N_{TAIL} from nucleocapsids renders them more rigid and less prone to fragmentation [89]. The plasticity of N_{TAIL} and its effect upon nucleocapsid structure can also explain the dramatic changes in nucleocapsid morphology of SV5, SeV, or VSV that is induced by selective proteolysis or changes in salt concentration [62]. Data from Longhi et al. also shows that binding of P (i.e., the X domain, or XD) to N_{TAIL} results in an unstructured-to-structured transition in the latter, a phenomenon known as "induced folding" [14, 15, 69] (see also Chapter 1). Negative stain TEM shows that this interaction results in nucleocapsid helical unwinding and enhanced exposure of the RNA genome (Bhella and Longhi, unpublished observations). Thus, the P-induced changes in N_{TAIL} conformation may serve a polymerase cofactor function that is in addition to the role of P as a tether between L and N. We propose that hsp72 can perform a similar role as P in inducing nucleocapsid conformational changes, since negative stain TEM of hsp72-CDV light nucleocapsid complexes revealed similar evidence of helical unwinding relative to nucleocapsid lacking hsp72 [107]. This evidence includes hsp72-associated increases in nucleocapsid outer and inner core diameters, decreased subunit resolution, an increased tendency to break into subunit rings, and increased susceptibility of the RNA genome to nuclease degradation (Figure 5).

Hsp72-mediated light nucleocapsid formation likely represents a destabilized structure, based upon the lower polymerase activity associated with these particles relative to dense nucleocapsid containing hsp72 and the propensity for nuclease degradation of the genome. The difference in density likely reflects diminished compaction of helical turns, resulting in an altered hydration state of the core particle. This change in hydration state would likely contribute more to the lower buoyant density than would the increased protein:RNA ratio associated with the addition of hsp72. The enhanced polymerase activity associated with dense nucleocapsid containing hsp72 suggests that RNA exposure can be enhanced without global helical destabilization, reflecting a lower magnitude or even regional hsp72 binding to N protein. This structural model explains the biphasic dose response of transcription to hsp72 supplementation. Low levels of hsp72 would stimulate activity in dense nucleocapsid particles. Increasing complex formation between hsp72 and dense nucleocapsid would further increase activity, but would also be accompanied by the formation of a light nucleocapsid by-product reflecting excessive helical destabilization. The proportion of dense to light nucleocapsid formation would determine whether the response to further increases in hsp72 levels would be stimulatory (and the degree of that stimulatory response) or inhibitory. Similar hsp72-induced changes in nucleocapsid density have been demonstrated for MeV [150], although characterization of ultrastructure and enzymatic activity of specific MeV nucleocapsid-hsp72 complexes has not been performed.

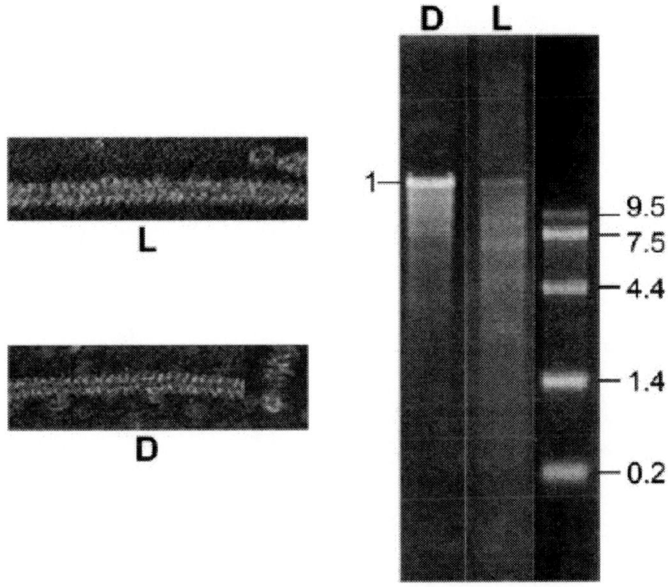

Figure 5. Ultrastructural differences between hsp72-dependent light nucleocapsid (L) and dense nucleocapsid (D) of CDV suggest that hsp72 promotes nucleocapsid helical unwinding and enhanced exposure of the genomic RNA. Negative stain transmission electron microscopy shows increased light nucleocapsid inner and outer core diameters and decreased subunit resolution compared to dense nucleocapsid (left). Agarose gel electrophoresis and ethidium bromide staining illustrates the susceptibility of genomic RNA from light nucleocapsid to nuclease degradation, in contrast to the protection of the 15 kilobase genome in dense nucleocapsid (right). The sizes of migratory standards are indicated in kilodaltons. Dense nucleocapsids examined in these experiments were devoid of hsp72.

Correlations between nucleocapsid structural variants and biological activity have long been noted, although mechanistic links were never established. The in situ ultrastructural appearance of CDV light nucleocapsid was identified based upon correlation of nucleocapsid diameters measured in situ and those measured from gradient purified preparations. These light nucleocapsid particles correspond to CDV and MeV variants described in literature as "granular", and granular nucleocapsids are associated with lytic infection [76, 125]. In contrast, the appearance of nucleocapsid that is devoid of hsp72 corresponds to the smooth variant associated with persistent infections [125]. According to the model presented above, granular nucleocapsid formation would be a by-product of the more productive interaction between hsp72 and smooth nucleocapsid, whereas exclusive formation of smooth nucleocapsid would suggest only a low level of hsp72 functional interaction. Modulation of viral transcription and genome replication is a well-recognized basis for interconversion between lytic and persistent modes of replication, including the heat shock-induced reactivation of stable persistent CDV infection [109]. Moreover, the connection between hsp72-nucleocapsid complex formation and a lytic infection phenotype is established for both MeV and CDV.

The structural nucleocapsid changes induced by hsp72 could explain both increased transcription and genome replication, but it is also possible that the underlying mechanisms are distinct. It is possible that hsp72 binding to N protein induces more than one nucleocapsid conformation, one that favors transcription and another that favors genome replication.

Transcription is initiated at a single promoter near the 3' end of the genome, using a start-stop mechanism to produce fully processed transcripts for each of the viral proteins (reviewed by Lamb and Kolakofsky [80]). Genome replication also begins with the production of plus stranded RNA, a replicative intermediate that serves as template for the production of negative strand RNA. Genome replication is initiated from promoters at the genomic termini that are composed of two discontinuous elements that combine to form a functional unit when juxtaposed by successive helix turns [145]. One model for the switch between genomic replication and transcription thus involves variation in nucleocapsid helical conformation. Changes in the number of N protein monomers per turn or "twist" affects the number of nucleotides per turn that, in turn, can disrupt the genomic promoter to enhance transcription or vice versa. Whether hsp72 influences nucleocapsid morphology in a way that can favor transcription or genome replication awaits structural analysis of nucleocapsid variants that can be assigned specific functional states. It is also possible that separable effects on transcription and genome replication reflect the existence of a distinct viral replicase and transcriptase, where hsp72 is differentially involved in the formation or activity of these distinct polymerase complexes. The existence of discrete RNA polymerase complexes have been documented for VSV, one being a replicase composed of N, P, and L proteins, and the other a transcriptase incorporating host factors that include hsp60 [122]. A functional role for the hsp60 remains to be established. A *Morbillivirus* transcriptase incorporating hsp72 could focus the nucleocapsid helical destabilizing activities of hsp72 to the area where such activity is needed - in the vicinity of the viral polymerase. These determinants of transcription versus genome replication are not mutually exclusive since structural variation in the nucleocapsid may determine whether the ribonucleoprotein template supports the formation/activity of a viral replicase or transcriptase.

The ability of hsp72 to enhance genome replication could occur via a more indirect mechanism, namely by facilitating encapsidation of the nascent genome formed by the viral replicase. Genome replication occurs only in the presence of soluble N (N^0) that is required for encapsidation of the newly formed strand (see Chapter 1). One role of the viral P protein is to bind N^0, maintaining its solubility and inducing a conformation that favors initiation of encapsidation on viral genomic termini versus promiscuous encapsidation of any RNA. The amino terminal disordered domain of P binds N_{CORE} and the ordered C-terminal domain of P binds N_{TAIL}, although the relative contributions of these distinct interactions to the fidelity of encapsidation remain to be shown. We tested the hypothesis that hsp72 may similarly support encapsidation reactions. Edmonston MeV minigenome RNA was synthesized in vitro and added to an in vitro transcription/translation system. This system supports T7-mediated expression of viral N and P proteins or hsp72 from plasmid vectors. Minigenomic RNA contained the CAT coding sequence flanked by the viral 5' and 3' non-coding genomic termini that include 5' signals for encapsidation. Once encapsidated, genomic RNA becomes resistant to nuclease degradation, allowing nuclease resistance to be used as a measure of encapsidation. Northern blot analysis of genomic RNA showed that both hsp72 and P enhance the specific encapsidation of genomic RNA without influencing the levels of N protein expression (Figure 6). Addition of N without hsp72 or P did not confer protection, indicating that specific encapsidation requires the presence of either P or hsp72. Protection was not observed following expression of either hsp72 or P alone, indicating the requirement of N for protection. These findings illustrate overlap in the function of hsp72 and P in supporting genomic encapsidation.

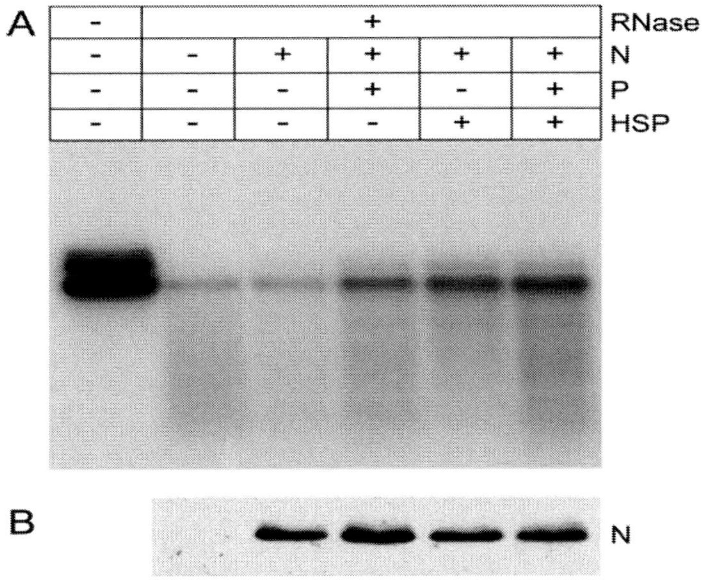

Figure 6. Encapsidation of MeV minigenomic RNA by N protein can be facilitated by hsp72, an effect otherwise mediated by the viral P protein. A MeV minigenomic RNA was transcribed in vitro and added to a reticulocyte lysate supporting plasmid-based expression of N, P, and/or hsp72 protein. (A) Reaction products were digested with S7 nuclease. Northern blot analysis of the minigenome RNA shows nuclease protection when N is co-expressed with hsp72 or P but not when N is expressed alone. (B) Western blot analysis shows that protection is not correlated with differences in the level of N protein expression.

2.4. HSP72 Binding to the Nucleoprotein

The definitive linkage between hsp72-nucleocapsid binding and functional consequences relies on a precise characterization of binding targets. Intrinsically disordered protein domains such as N_{TAIL} represent attractive binding targets for HSPs due to the accessibility of short hydrophobic patches that can be contained within their structure. The observation that bacterial expressed MeV N_{TAIL} co-purifies with the bacterial homologue of mammalian HSP70 (i.e., DnaK) supports the HSP-binding potential of this viral protein domain (Longhi et al., unpublished observation). Direct evidence that N_{TAIL} mediates hsp72 binding to *Morbillivirus* nucleocapsids came from the analysis of the interaction between purified hsp72 and purified Edmonston MeV nucleocapsid where N_{TAIL} was removed by selective proteolysis with V8 protease [163]. The highly conserved amino terminal N_{CORE} drives helix formation and is resistant to proteolysis when incorporated into nucleocapsid. The susceptibility of N_{TAIL} to proteolysis reflects not only is surface exposure but also an intrinsic property of disordered protein domains. Binding reactions were measured using surface plasmon resonance (SPR) technology, where the hsp72 is immobilized on sensors and the nucleocapsid passed over the surface in solution (i.e., the reaction analyte), with both association and dissociation phases of the reactions being monitored in real time. Nucleocapsid particles are sheared to generate a relatively short and uniform particle length, thereby facilitating this type of analysis. Binding reactions for multiple concentrations of nucleocapsid are examined, allowing computation of reaction rate and equilibrium constants, with the equilibrium dissociation constant (K_D) being the measure of binding affinity. Binding affinity between nucleocapsid and hsp72 (K_D = 16 nM) is reduced approximately 25-fold when N_{TAIL} is removed by selective proteolysis. Purified recombinant N_{TAIL} from Edmonston MeV was also used as analyte and found to bind hsp72 with an affinity (and rate constants) that are comparable to that between hsp72 and nucleocapsid (i.e., K_D = 11 nM) (Table 1) [162].

Putative hsp72 binding sites within N_{TAIL} were identified by searching for conserved regions of hydrophobicity. Although the sequence of N_{TAIL} is hypervariable, there are three regions ≤ 20 amino acids in length that are conserved and enriched in amino acids with hydrophobic side-groups. These regions are referred to as Box-1 (N 401-420), Box-2 (N 489-506), and Box-3 (N 517-525). Valentin and Longhi identified Box-1 as the binding determinant for the nucleoprotein receptor (NR) [78] (see also Chapter 5). NR mediates binding of extracellular MeV N protein that is released from necrotic cells to the surface of thymic epithelial cells and T cells. NR binding by N results in cell cycle arrest that may be a basis for MeV-induced immunosuppression [79]. Longhi et al. identified Box-2 as a binding determinant for the X domain of P [14, 15].

Table 1. SPR analysis of binding reactions between hsp72 (ligand) and Ed-MeV nucleocapsids, N_{TAIL}, or peptides representing N_{TAIL} motifs (analyte). Binding affinities are reflected in the equilibrium dissociation constants (K_D). Shaded panels emphasize the similarity in hsp72 binding affinity for nucleocapsid, N_{TAIL} protein, and Box-2 peptide

Peptide Analytes	k_a(1/Ms)	k_d (1/s)	K_D (M)
Box 1 peptide	0.2	$5.6x10^{-3}$	24 mM
Box 2 peptide	$4.1x10^{2}$	$1.0x10^{-5}$	24 nM
Box 3 peptide Ed N	$1.4x10^{2}$	$4.9x10^{-4}$	3.6 μM
Box 3 peptide Ed N-522D	7.0	$5.5x10^{-3}$	0.8 mM
Box 3 peptide CDV	27	$1.0x10^{-4}$	6.0 μM
Ed nucleocapsid	$6.4x10^{2}$	$1.0x10^{-5}$	16 nM
Ed N_{TAIL}	$8.8x10^{2}$	$1.0x10^{-5}$	11 nM
Ed $N_{TAIL522D}$	$8.9x10^{2}$	$1.0x10^{-5}$	11 nM
Ed $N_{TAIL\Delta2,3}$	6.0	$2.3x10^{-5}$	3.8 μM
Ed nucleocapsids ΔN_{TAIL}	23	$1.0x10^{-5}$	0.4 μM

An algorithm used to predict HSP70 target sequences identified Box-1, Box-2, and Box-3 as putative hsp72 binding sites. The algorithm examines sequence composition within linear stretches of 8-9 amino acids. The presence of at least two non-contiguous hydrophobic and/or aromatic side groups (F, W, Y, I, V, L) and a paucity of acidic residues (E, D) predict

HSP70 binding, whereas internal basic residues (R, K) confer selectivity for hsp72 [43, 51]. Box-3 sequence is unique in that it is highly conserved between MeV and CDV, and we demonstrated the hsp72 binding capacity of Box-3 using SPR analysis [163, 164] (Table 1). Box-3 synthetic peptides based upon Edmonston MeV (Ed-MeV) and Onderstepoort-CDV (Ond-CDV) bind hsp72 with a comparable and physiologically relevant affinity (3.6 and 6 μM, respectively), values consistent with activity control reactions involving p53 and hsp72 [42, 43]. Naturally occurring sequence polymorphisms of Box-3 were identified for MeV when examining 174 isolates representative of all genotypes [77]. Hydrophobic and basic residues were invariant in 98% of the strains. However, 94% of wild type viruses differed from Ed-MeV in that they exhibited an increased content of acidic residues due to an N to D substitution at position 522, a change predicted to diminish hsp72 binding affinity. SPR analysis confirmed this prediction, showing an approximately 800-fold reduction in binding affinity for N-522D peptide. All members of the group A genotype of MV preserve the Box-3 sequences with a higher affinity for hsp72. SPR analysis ruled out binding between hsp72 and Box-1 based upon the low affinity for Box-1 peptide (K_D = 24 mM) and the fact that N_{TAIL} binding to hsp72 is not influenced by the deletion of Box-1. These finding are supported by the inability to disrupt NR/N_{TAIL} interactions through competitive inhibition by hsp72 (Valentin et al., unpublished observation; see also Chapter 5).

Although hsp72 can directly bind Box-3, the low affinity for Box-3 indicates that this is not the primary driving force for hsp72-nucleocapsid interaction. This conclusion is based upon the fact that N_{TAIL} constructs incorporating the 522D Box-3 variant bind hsp72 with an affinity that is comparable to the parent N_{TAIL} [162] (Table 1). Secondly, nucleocapsid of recombinant infectious MeV incorporating the 522D Box-3 variant retains the ability to form complexes with hsp72 that can be isolated on CsCl gradients [162]. Instead, it is Box-2 that is the driving force for hsp72-nucleocapsid interaction (Figure 7) [162]. Hsp72 binds Box-2 peptide with an affinity that is two orders of magnitude greater than the affinity for Box-3 (K_D = 24 nM) (Table 1). The affinity of hsp72 for Box-2 peptide, Ed-MeV nucleocapsids, and purified recombinant N_{TAIL} are all comparable (Table 1). N_{TAIL} in which Box-2 and Box-3 are deleted ($N_{TAIL\Delta 2,3}$) exhibits a marked reduction in hsp72 binding affinity, similar to the loss of nucleocapsid-hsp72 binding affinity when N_{TAIL} is removed by selective proteolysis (ΔN_{TAIL}) (Table 1) [163]. Unlike Box-3, the putative amino acid determinants of hsp72 binding to Box-2 (i.e., the amino acids with hydrophobic or charged side groups) are well-conserved amongst MeV isolates.

Results of hsp72/N protein co-immunoprecipitation experiments support the importance of Box-2 to hsp72 binding [162]. In this approach, Hep2 cells were transfected with plasmids supporting T7-mediated expression of hsp72 and either N, N-522D, N lacking Box-3 (NΔ3), or N lacking both Box-2 and 3 (NΔ2,3), with T7 polymerase expressed from a recombinant replication defective vaccinia virus (MVA). Antibody against hsp72 pulled down N, N-522D, and NΔ3 but not NΔ2,3. No N construct was recovered in the absence of hsp72 overexpression and results were confirmed using a reciprocal approach (i.e., use of an anti-N monoclonal antibody to identify complex formation with hsp72).

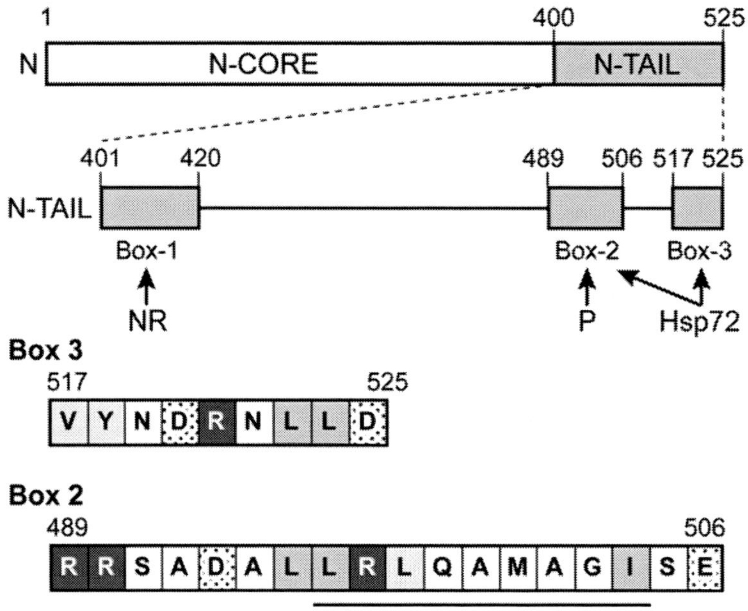

Figure 7. Ed MeV NTAIL binding targets of hsp72 and the P protein X domain. Box-2 is a primary recognition motif for both hsp72 and P. Although Box-3 stabilizes P/NTAIL complexes, peptide binding affinity between P and Box-3 suggest that primary contacts do not occur. In contrast, hsp72 can directly bind Box-3 of Ed MeV with low and physiologically relevant affinity. The sequences of Box-2 and Box-3 are shown, emphasizing probable determinants of hsp72 binding. Hydrophobic (light grey) and basic (dark grey) residues favor hsp72 binding whereas acidic residues (stippled) discourage binding. The putative nominal hsp72 binding region in Box-2 is underlined.

These results rule out contributions by the N protein amino terminal assembly domain (N_{CORE}) in the reversible association that characterizes purified hsp72-nucleocapsid complexes but they do not rule out possible hsp72/N_{CORE} interactions that could mediate nucleocapsid assembly. N_{CORE} drives the formation of nucleocapsids-like particles in cells expressing N [72], and our results suggest that N-N interaction driven by N_{CORE} can ultimately out-compete any hsp72-N binding that may target this region. Such behavior is typical of chaperone functions where an HSP maintains the solubility of a hydrophobic protein monomer and facilitates its orderly incorporation into the multimeric complexes that the monomer forms. Recovery of hsp72 complexes with NΔ3 was less than with parent N, suggesting that Box-3 may stabilize hsp72/N_{TAIL} binding that is otherwise driven by Box-2. This situation would be similar to that observed for binding of the P protein XD to N_{TAIL}.

SPR analysis of binding reactions between the X domain of the P protein (XD) and N_{TAIL} indicate many similarities with hsp72 binding reactions. First, XD binds Box-2 peptide and N_{TAIL} with a high affinity that is within the range of that observed for hsp72 [15]. XD binding to Box-2 peptide is characterized by a K_D of 20 nM and binding to N_{TAIL} by a K_D of 81 nM. The validity of these SPR-derived values was confirmed by fluorescence spectroscopy using W-substituted N_{TAIL} (F518W N_{TAIL}), where a K_D of 133 nM is calculated for XD/ N_{TAIL} binding [15]. SPR analysis confirms dual recognition of Box-2 by hsp72 and P by showing that hsp72 can competitively inhibit XD/N_{TAIL} binding reactions [162]. A negative control for these competitions was purified recombinant hsp72 peptide binding domain (PBD, amino acids 381-541), generated from the same construct used to produce the full length molecule. Failure of the PBD alone to interfere with N_{TAIL}/XD binding indicated the importance of the hsp72 nucleotide binding domain in supporting high affinity hsp72/N_{TAIL} interactions. Competition between hsp72 and viral P protein is likely to occur under physiological conditions when one considers the fact that the concentration of soluble hsp72 that is found in preconditioned Vero cells is estimated to be 85 nM [110]. Another similarity between XD and hsp72 is that neither recognizes Box-1 to a significant degree. The K_D describing the XD/Box-1 peptide binding reaction is 9 mM and XD binding to N_{TAIL} is not affected by the deletion of Box-1.

The one difference between XD and hsp72 is that XD does not exhibit a significant binding affinity for Box-3 peptide [15]. The K_D for binding reactions between XD and Box-3 peptide is approximately 1 mM which is outside of the physiological range. This is contrasted to the low but physiological relevant K_D of 3.6 μM for hsp72. The fact that XD has a low affinity for direct Box-3 binding does not, however, indicate that XD/Box-3 interactions are unimportant. To the contrary, it is apparent that Box-3 is needed to stabilize XD/ N_{TAIL} complexes based upon the observation that N_{TAIL} lacking Box-3 exhibits a 1000-fold reduction in binding affinity for XD relative to that of the parent N_{TAIL} molecule. This reduction is similar to that observed when both Box-2 and Box-3 are deleted from N_{TAIL}. It is likely that XD-induced folding of N_{TAIL} allows Box-3 to make secondary contacts with the ligand of Box-2, with these additional contacts providing the stabilization of the XD/N_{TAIL} complex. The latter is supported by structural analysis of XD/N_{TAIL} complexes using small angle X-ray scattering and heteronuclear magnetic resonance [15]. It is unknown if hsp72 induces similar secondary contacts with Box-3 following Box-2 binding. Although our data suggest a contribution of Box-3 to stabilization of hsp72/N_{TAIL} complexes, we have no evidence to suggest that the magnitude of this effect is as great as it is for the P protein.

Similarities between hsp72 and P with regard to N_{TAIL} binding suggest yet another mechanism by which hsp72 might enhance viral polymerase activity, namely to increase polymerase processivity. This processivity would be enhanced by relaxing the tight association between the polymerase and its nucleocapsid template, and this would be achieved by hsp72's ability to competitively inhibiting binding of P to Box-2, or by binding Box-3 to remove its stabilizing influence on P/N_{TAIL} complexes.

2.5. Functional Consequences of HSP72 Binding to N_{tail}

Minireplicon reporter gene expression is an efficient way to determine the influence of both viral and cellular proteins on viral transcription and genome replication, although the assay does not readily distinguish between these two polymerase functions. Target cells (i.e., HeLa, Hep2, or Vero) are transfected with plasmids that support production of MeV N, P and L messenger RNA and MeV genomic RNA from the T7 promoter. The source of T7 polymerase is MVA. Coding regions of the MeV or CDV genome are replaced with a reporter gene such as chloramphenicol acetyl transferase (CAT) while preserving viral genomic termini that are promoters for replication, encapsidation, and transcription. The template for transcription/replication (i.e., the encapsidated minigenome) is assembled within the cell, relying entirely upon plasmid DNA as a source of N, P, and L. The roles of Box-2 and Box-3 in providing a template for P during transcription and genome replication is illustrated by using Ed-MeV N protein truncation mutants in this system [163]. We showed that N protein lacking Box-3 or sequences between Box-2 and 3 support increased reporter gene expression relative to parent protein, supporting the model whereby Box-3 acts as a clamp to hinder polymerase processivity. Truncations involving Box-2 abrogate viral gene expression, consistent with the role of Box-2 as a primary recognition motif for P in support of transcription and genome replication (Figure 8).

MeV minireplicon reporter gene expression is stimulated by hsp72 that is produced by either transient heat shock or by cotransfection with an hsp72 expression plasmid [116, 163]. For plasmid transfections to express hsp72, recombinant human hsp72 is tagged with a VSV-G epitope so that endogenous and plasmid-expressed hsp72 can be distinguished by Western blot analysis. The plasmid-based approach provides a more controllable method of increasing hsp72 levels, allowing the dosage-dependent relationship between hsp72 and minireplicon reporter gene expression to be established. Hsp72 enhances MeV reporter gene expression by 3-10 fold, although excessive supplementation of hsp72 disrupts viral gene expression, similar to the biphasic hsp72 dose response observed in cell-free nucleocapsid transcriptional assays. A similar hsp72-mediated stimulation of minireplicon reporter gene expression has been shown for CDV (Oglesbee et al., unpublished observation).

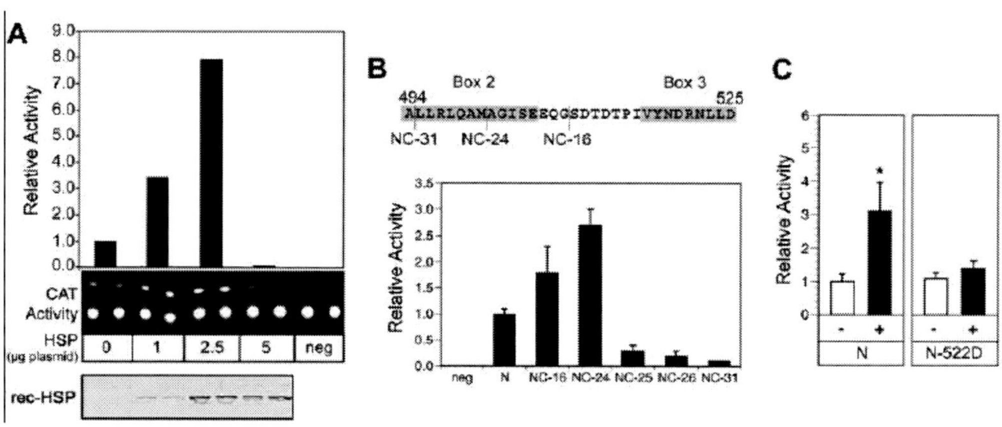

Figure 8. Influence of N protein mutation and/or hsp72 supplementation on MeV minireplicon reporter gene (i.e., CAT) expression [163]. CAT activity is expressed relative to a value of 1.0 for minireplicons using Ed-MeV N protein as template. (A) hsp72 supplementation increases reporter gene expression in a biphasic dosage-dependent manner. Lower levels of hsp72 stimulate whereas high levels suppress gene expression. Negative control reactions omit the viral polymerase. Levels of the recombinant VSV-G tagged hsp72 are demonstrated by Western blot analysis of cell lysate. (B) Sequential C-terminal amino acid truncations of N (i.e., NC – n) cause increased basal reporter gene expression until they involve residues within Box-2, the binding site for P protein. Deletions within Box-2 abrogate viral gene expression. (C) N protein containing Box-3 sequence from wild type virus (N-522D) supports basal reporter gene expression that is identical to Ed-MeV N but does not support the increase otherwise mediated by hsp72 supplementation (1 µg plasmid).

Mutations of the N protein providing the template for minireplicon reporter gene expression revealed two functional attributes of Ed MeV Box-3 [163]. First, C-terminal deletions involving Box-3 ± the region between Box 3 and 2 resulted in loss of hsp72 responsiveness in addition to the increase in basal transcription/genomic replication. Secondly, the wild type variant of Box-3 (i.e., N-522D) supports a basal level of reporter gene expression that is identical to that of the Edmonston vaccine virus, yet loses the responsiveness to elevated hsp72 levels. These findings support the role of Box-3 as a regulatory domain that represses transcription/genome replication. Destabilization of the complex between Box-2 and P protein, and the associated increase in polymerase processivity, can be mediated by either deleting Box-3 or by neutralizing its effect through hsp72 interaction.

Vaccine versus wild type Box-3 variants were tested in the context of infectious virus in order to better discriminate between effects of hsp72 on transcription versus genome replication [24, 162]. The rescue system for infectious virus is the same as the minireplicons except that full-length antigenomic RNA is expressed from plasmid instead of the minigenome. The plus strand antigenome is encapsidated by N, which then serves as template for the production of minus strand genomes that initiate the replicative cycle. For the N-522D variant, the mutation is created in the genomic backbone so that N protein expressed during subsequent replication cycles is N-522D. Differences in basal replication parameters between virus expressing N or N-522D were not observed on Vero or murine neuroblastoma (N2A) cells. Viruses were then propagated on hsp72 overexpressing (i.e., preconditioned) and nonconditioned control cell lines.

Northern blot analysis of viral RNA showed that the N-522D variant of Box-3 selectively attenuates hsp72 stimulation of viral transcription in Vero cells preconditioned by hyperthermic treatment, or human astrocytoma and murine neuroblastoma cell lines stably transfected to constitutively overexpress human recombinant hsp72 [25, 162]. Viral genome levels were increased by hsp72 in cells infected by both Ed-MeV and the N-522D variant, indicating that hsp72-dependent stimulation of transcription and genome replication are separable events. Increased genome levels do serve as template for (secondary) transcription, this influencing the degree to which hsp72-dependent increases in transcript levels are attenuated. These results indicate that the minireplicons were measuring the stimulatory effects of hp72 upon transcription but not genome replication. That hsp72 may enhance genome replication for infectious virus but not minireplicons may reflect the 18-fold greater genome length of the former, placing greater demands upon host cofactors such as hsp72 in supporting genome replication (e.g., by enhancing polymerase processivity or encapsidation of nascent genomic RNA) [162]. In other words, low levels of hsp72 are rate limiting for genome replication of infectious virus but not for minireplicons.

Hsp72-dependent increases in CPE and cell-free infectious viral progeny release are attenuated in the N-522D variant virus, supporting the role of hsp72-dependent increases in viral transcription as the basis for these changes in infection phenotype [25, 162]. CPE is readily quantified by measuring plaque area in Vero cells, a measure of the degree of syncytia formation on monolayers that is mediated by expression of the viral fusion (F) and hemagglutinin (H) proteins. Infection of control Vero monolayers by the N-522D variant results in a unimodal population of small plaques that are similar to those formed by the parental Ed-MeV, whereas infection of hyperthermic preconditioned Vero cells results in the emergence of large plaques for Ed-MeV but not the N-522D variant (Figure 9).

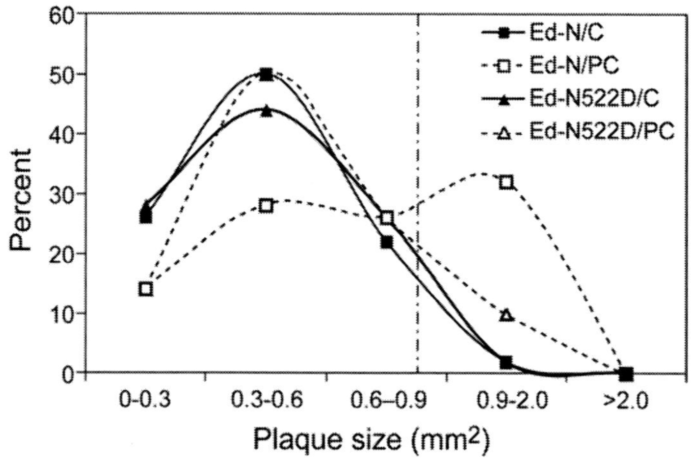

Figure 9. Plaque area analysis shows that parent Ed-MeV (Ed-N) forms large and small plaques on preconditioned (PC) Vero cells versus a unimodal distribution of small plaques on control cells (C). Large plaque forming ability is attenuated in virus bearing the N-522D mutation.

The impact upon cell-free progeny release was most clearly demonstrated in neuroblastoma cells. The kinetics and magnitude of release were identical for Ed-MeV and the N-522D variant on vector transfected control cell lines. Significant increases in progeny release were observed on transgenic hsp72 overexpressing cells at 72 and 84 h PI, but only for the parent Ed-MeV, where the increase over controls was greater than one order of magnitude for both time points [25].

Similar experiments recently performed with wild type MeV isolates confirm the inability of 522D Box-3 motifs to support hsp72-dependent stimulation of transcription and CPE, and the more general ability of hsp72 to stimulate viral genomic replication (Oglesbee et al., unpublished observation). These experiments were performed using the ICB, WTF, and Bilthoven MeV strains to infect pre-conditioned or control Vero cells stably transfected to express CD150, the receptor for wild type MeV. In addition, the transcriptional responsiveness of 522N Box-3 motifs was confirmed using the Schwartz and Zagreb MeV vaccine strains. A role of Box-2 in hsp72-mediated stimulation of genome replication remains to be established, as does the possibility that hsp72 must interact with both Box-2 and Box-3 in order to stimulate transcription. Unraveling the contributions of Box-2 to hsp72-dependent transcription and genome replication will rely upon a mutational approach designed to disrupt hsp72 but not P binding. Feasibility of this approach is supported by the fact that binding targets discriminate between P and hsp72, based upon the observation that Box-3 peptide binds hsp72 with a ~300-fold greater affinity than XD. Preliminary SPR studies from our laboratory indicate that increasing the content of acidic side groups within the nominal hsp72 recognition motif in Box-2 (i.e., A502D and G503D amino acid substitutions) can reduce hsp72 binding affinity by 1,000-fold without affecting XD binding affinity.

2.6. A Model for the Structure-Function Relationship between HSP72 and MeV N Protein

The structure-function model of hsp72/N_{TAIL} interaction integrates our emerging view of P/N_{TAIL} interactions where P binds Box-2 to induce conformational changes in MeV N_{TAIL} that favor transcription and/or genomic replication. Secondary Box-3/P interactions stabilize the P/N_{TAIL} complex, hindering the processive movement of P along the ribonucleoprotein template. Hsp72 binds Box-3 in "responsive" viral strains to diminish the strength of P/N_{TAIL} interactions to levels that are more permissive to transcription. The selectivity for transcription versus genome replication in this model could be explained by selective incorporation of hsp72 into a transcriptase, analogous to the hsp60 component of the VSV transcriptase. The inability of the majority of MeV wild type isolates to support hsp72/Box-3 interactions suggests an advantage of transcriptional constraint in the face of elevated hsp72 within an infected human. Mixed infections of Ed-MeV and the N-522D variant in both culture systems and in the cotton rat model of airway infection show that fitness of the N-522D variant is reduced, supporting the existence of selectional pressures that assure circulation of viruses containing this motif within human populations [24].

Hsp72 may also enhance polymerase processivity by competitively inhibiting P binding to Box-2, an interaction that would be supported by all MeV strains. Characterization of gene expression by infectious viral variants or different MeV strains in preconditioned cells indicates that binding to Box-2 alone is not sufficient to influence transcription, although the possibility remains that this interaction is necessary for transcription, operating in conjunction with Box-3. Box-2 binding may also be responsible for the hsp72-dependent stimulation of genome replication that is observed in Ed-MeV, the N-522 variant, and the wild type MeV isolates that have thus far been examined. The effect may represent enhanced assembly or processivity of a viral replicase, nucleocapsid structural transitions that not only enhance exposure of the genomic RNA but also favor activation of the promoter for genome replication, or enhanced encapsidation of nascent genomic RNA. The effects may reflect hsp72-induced conformational changes in N_{TAIL} that are both common to and distinct from those caused by the P protein, the result being that hsp72 provides cofactor functions that are otherwise mediated by P.

3. Biological Significance of HSP72-Nucleocapsid Interactions

There is a tendency to assume that any variable stimulating viral gene expression, CPE, and progeny release should enhance viral virulence. Even small changes detected in an in vitro system could be clinically significant due to the amplification of these changes over the countless rounds of replication that would occur *in vivo* (i.e., within an animal host). However, we must consider the impact of enhanced viral gene expression on innate and adaptive immune responses that drive viral clearance and the potential for increased CPE to hinder viral spread in tissue compartments where infection is primarily cell-to-cell versus infection by cell-free infectious viral progeny. We must also acknowledge the role of hsp72 in promoting both innate and adaptive immune responses (see Oglesbee et al, 2002, for review [111]). Hsp72 can be expressed on the cell surface of virus infected cells, representing a target of lymphokine activated killer cells. Hsp72 that is released from necrotic cells can activate antigen presenting cells and stimulate release of proinflammatory cytokines, in addition to mediating the cross-presentation of bound antigen into MHC I presentation pathways. The ability of hsp72 to promote strong cellular immunity and clearance is illustrated by vaccine studies in mice. Expression of a viral structural protein of Japanese encephalitis virus or herpes simplex virus is protective against subsequent lethal viral challenge when co-expressed with hsp72, and the adjuvant effect is observed in both adolescent and neonatal mice [24, 114].

3.1. HSP72 Modulates MeV Neurovirulence

Our work has examined the effect of elevated hsp72 on Ed-MeV virulence using the mouse model of MeV encephalitis. The model is attractive since mice lack a febrile response following intracranial viral inoculation with Ed-MeV and they lack the low constitutive levels of hsp72 that can be observed in other animal species [22, 115], maximizing control over hsp72 expression levels that can be driven by transient hyperthermic treatment or transgenic expression. The hsp72 responsiveness of Ed-MeV is well defined in mouse neuroblastoma cells [24] and neurons are a primary target of infection in mouse models of Ed-MeV encephalitis [10, 101]. Neuronal infection is also relevant to MeV pathogenesis in its natural primate hosts, this being part of a multisystemic dissemination of virus following initial replication in lymphoid tissue. Invasion of the central nervous system (CNS) is common albeit clinically silent in the majority of cases [18, 73]. Electroencephalographic evidence for CNS invasion has been observed in patients following MeV vaccination [33] and in 50% of uncomplicated cases of naturally occurring MeV infection [18]. MeV N protein transcripts have been detected in brain of 18% of autopsy cases unassociated with MeV-induced neurological disease [73]. Less common forms of MeV CNS infection include measles inclusion body encephalitis (MIBE), an acute fulminate infection induced by either wild type or vaccine virus in immunocompromised patients, and subacute sclerosing panencephalitis (SSPE), a manifestation of viral persistence that occurs years following early childhood exposure to wild type MeV [39]. Unresolved are the viral and host determinants that allow persistence to become established and what (if any) relationship exists between these determinants and MIBE and SSPE.

In neonatal mice, MeV cell-mediated clearance is hindered by host-restricted low levels of viral gene expression, direct cell-to-cell transmission of virus, and constraints on antigen presentation that are inherent to the CNS [87]. Mouse strains can be defined as resistant or susceptible to MeV infection of brain based upon H-2 restricted differences in virus-specific CD4+ and CD8+ T cells responses that mediate viral clearance during the acute phase of infection [102, 103, 154, 155]. Our work shows that in a susceptible strain such as C57BL/6 (H-2^b), intracranial inoculation with Ed-MeV results in approximately 15% mortality during the acute phase (i.e., $\leq$ 28 d PI), with persistence in remaining animals documented up to 70 d PI [23]. This result is consistent with the low yet variable mortality rates of 0-30% that has been reported by other groups [123, 128]. Challenging neonatal H-2^b C57BL/6 mice via the intracranial route thus exploits both mouse age and strain-dependent susceptibilities to MeV infection [41, 55, 84, 101, 103] and hence eliminates the need for transgenic expression of human membrane cofactor protein (CD46) for use as an ancillary viral receptor or use of mouse brain adapted strains of MeV in order to promote brain infection [38, 83, 99, 123]. The low incidence of mortality and the high incidence of persistence are ideal for documenting hsp72-mediated effects that may either increase virulence or clearance. In a resistant strain such as BALB/c (H-2^d), intracranial inoculation with Ed-MeV results in no mortality and clearance in approximately 53% of the animals during the acute phase of infection [22]. Antibody depletion of T cells show that CD4+ cells mediate protection in resistant strains of mice [41]. Clinical and experimental data from humans also stress the importance of cell-mediated immunity in recovery from MeV infection, although their role in overcoming infection remains to be shown [54, 67, 148, 149].

Transgenic C57BL/6 mice were generated that constitutively overexpress hsp72 in neurons to determine the impact upon Ed MeV infection [23]. The highly inducible form of 70 kDa HSP is produced from the hsp70-A2 gene in both humans and mice, and they share 98.2% similarity at the amino acid level [12]. Human hsp72 was used as the transgene since this gene product had been extensively tested in the in vitro analysis of MeV and CDV infections, including the establishment of stably transfected murine neuroblastoma cell lines that constitutively overexpress hsp72 and support hsp72-dependent increases in MeV gene expression, CPE, and infectious progeny release [25, 151, 163]. The VSV G tagged construct previously employed in minireplicon experiments was again used in order to distinguish transgene from endogenous hsp72 [163]. This construct was inserted into an L7 cassette that stabilizes transcripts in neurons [161], and this in turn was placed downstream of the rat neuron specific enolase promoter [70]. Neuron-specific expression of VSV G/hsp72 was shown by immunohistochemistry, with additional support of the neural tissue specific expression supported by Western blot analysis of tissue total protein extracts and RT-PCR of total RNA.

Using this system, we showed that hsp72-dependent increases in MeV transcription significantly enhance the neurovirulence of Ed-MeV [23]. Non-transgenic and transgenic neonatal mice received intracranial inoculations with $4x10^4$ PFU Ed-MeV. Viral RNA burden was measured in total brain RNA by SYBR green real time RT-PCR during the acute phase of the infection (2-4 weeks PI) and at the termination of the study (70 d PI). Mean viral RNA burden was approximately two orders of magnitude higher in transgenic versus wild type mice at 2-4 weeks PI and this was associated with an increase in virus-induced mortality that was restricted to the acute phase of infection, reducing median survival time from 70+ days in wild type mice to 17 days in transgenic mice. Overall mortality in the transgenic group was 65% compared to 13% for non-transgenic mice of the parental strain. Increases in viral RNA burden and mortality were associated with enhanced CPE in brain during the acute phase of infection, characterized by neuronal necrosis, neuronal cytoplasmic and intranuclear inclusion body formation, and the unique formation of neuronal syncytia in the CA1-3 layers of the hippocampus (Figure 10). Inflammatory changes were conspicuously absent. Northern blot analysis of total RNA from animals representing the mean viral burden at 2-4 weeks PI showed that the increased CPE is associated with dramatic increases in the expression of N and H transcripts, with H expression known to be a rate-limiting factor for syncytia formation [152] (Figure 10).

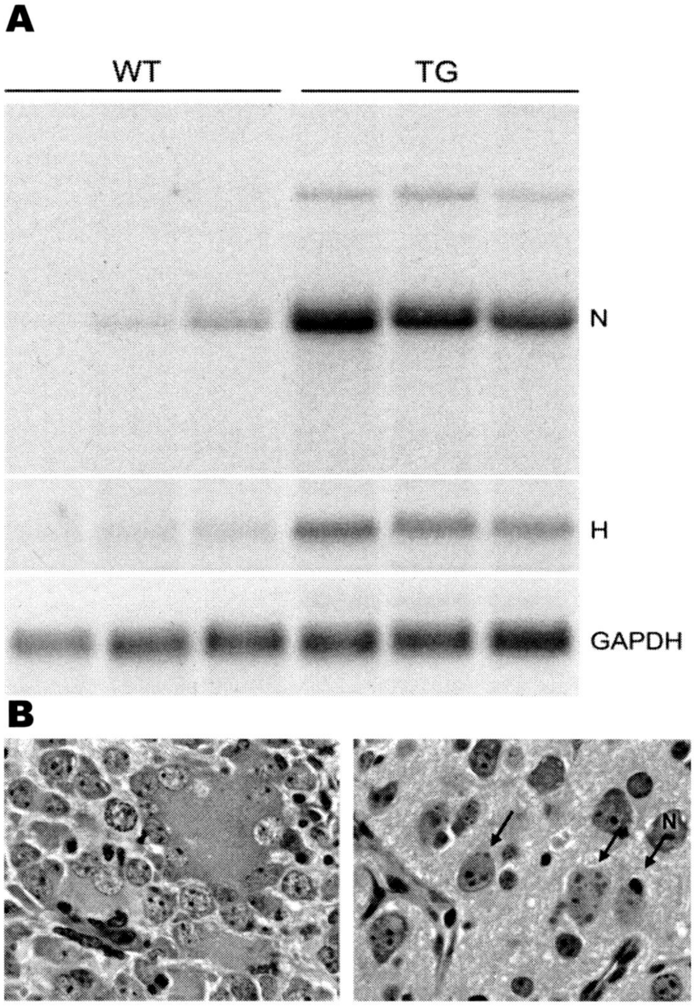

Figure 10. Elevated neuronal hsp72 levels in brains of neonatal C57BL/6 mice support increased Ed-MeV transcription and cytopathic effect [23]. Selective neuronal transgenic expression of hsp72 is driven by a neuron specific enolase promoter. Intracranial inoculation of transgenic and non-transgenic mice of the parental strain (wild type, or WT) used 4x104 PFU Ed-MeV. The mean brain viral RNA burden was approximately two orders of magnitude greater in transgenic relative to non-transgenic mice based upon real time RT-PCR of brain total RNA. (A) Northern blot analysis of samples representative of the mean viral RNA burden show increases in viral transcripts that are both proximal (N) and distal (H) to the genomic promoter for transcription. (B) The increased viral RNA burden in transgenic mice supported the unique formation of neuronal syncytia in the hippocampus (left), reflecting increased expression of viral membrane glycoproteins and the close proximity of neuronal cell bodies in this tissue compartment. The incidence of animals having neurons with cytoplasmic inclusion bodies (arrows) and neurons undergoing degeneration and necrosis (N) was increased from 30% to 100% in transgenic mice relative to non-transgenic mice.

Immuno-histochemistry confirmed expression of H glycoprotein in syncytia, reflecting mean viral burdens $\geq$ 1x10^8 copies of viral RNA/250 ng total brain RNA. *In situ* hybridization of viral N gene-specific RNA also showed that the virus-induced CPE was directly attributable to virus infection and not indirect mechanisms such as excitotoxicity. All of these tissue changes, including the formation of syncytia, are diagnostic features of MIBE in humans [39, 40, 44].

Infection of the hsp72 overexpressing transgenic C57BL/6 mice with the N-522D MeV variant showed that hsp72-dependent increases in viral virulence were due to the hsp72/N_{TAIL} interactions that result in increased transcription [23]. Brain viral RNA burden during the acute phase of infection was not influenced by hsp72 for the N-522D virus, being the same for transgenic mice and non-transgenic mice of the parental strain. Accordingly, there were no significant differences in viral CPE, which was minimal in both challenge groups, and the level of mortality was similarly low. Brain viral RNA burdens were similar and low during the chronic phase of infection for both transgenic and non-transgenic mice challenged with either Ed-MeV or the N-522D variant, suggesting that viral persistence is not influenced by hsp72/N_{TAIL} interactions in animals that survive the acute phase of infection.

Experiments are in progress to determine if mouse strain-dependent differences in cell mediated immune responses to MeV infection may determine the clinical manifestations of hsp72-dependent increases in MV gene expression. This is being addressed by back-crossing the C57BL/6 (H-2^b) hsp72 overexpressing mice against C57BL/10.D2 (H-2^d) mice. Preliminary studies show no mortality in transgenic H-2^d mice challenged with 4x10^4 PFU Ed-MeV compared to 32% mortality that was observed in non-transgenic mice of the parental strain. This result is similar to previous studies using transient whole body hyperthermia to elevate hsp72 in brains of neonatal Balb/c (H-2^d) mice prior to infection with Ed-MeV [22]. Preconditioning was achieved with a 41°C 30 min whole body hyperthermia. Induction of hsp72 was documented at 6 h post treatment by Western blot analysis of brain total protein. Tissue levels of hsp72 remain elevated for 24-48 h post treatment, reflecting the prolonged half-life of the protein [115]. The treatment is without significant pleiotropic effects, based upon the observation in other species that hsp72 can be induced in neurons by hyperthermia without morphological or immunohistochemical evidence of cellular injury, activation of cytokine cascades, induction of MHC, or derangements in neurological function reflected in altered tissue neurotransmitter levels or electrophysiological responses [36, 108]. Intracerebral inoculation with 4x10^4 PFU Ed-MeV was performed in preconditioned mice (at 6h post-treatment) and age-matched nonconditioned controls. Forty seven percent of the control animals supported a persistent cytopathic infection at $\geq$ 3 weeks PI based upon real time RT-PCR of total brain RNA. CPE in the infected brains was proportionate to viral RNA burden. In contrast, infected preconditioned mice lacked significant CPE and clearance was demonstrated in 95% of the animals. Analysis of shorter post infection intervals showed that clearance is first evident in both groups at 14 d PI. The temporal onset and progression of clearance was correlated to splenocyte blastogenic responsiveness to purified MeV antigen but not the production of MeV-specific antibody. Collectively, these results suggest that hsp72-dependent stimulation of Ed-MeV gene expression in an immunocompetent host (i.e., a host capable of mounting effective cell-mediated antiviral immune responses) is host protective.

3.2. A Comprehensive Model for MeV-HSP72 Interaction in the Infected Host

We have used these data to construct a more comprehensive model of MV N_{TAIL}/hsp72 interaction in brain of a natural host (Figure 11). In this model, viral infection induces hsp72 at levels that are sufficient to support basal viral transcription and genome replication, allowing for a persistent mode of replication that does not signal immune recognition of infected cells. These low concentrations of hsp72 are sufficient for structural and functional interactions between hsp72 and N_{TAIL} Box-2, given the high affinity exhibited between these binding partners. Further elevations in hsp72 levels would be associated with febrile responses. The low affinity interactions between N_{TAIL} Box-3 of hsp72 responsive virus would now be observed in addition to increased Box-2 binding, the result being an increase in both genome replication and transcription. Within an immunocompetent host, the hsp72-dependent increases in MeV transcription would facilitate immune clearance by overcoming host restricted low-levels of viral gene expression [22, 111].

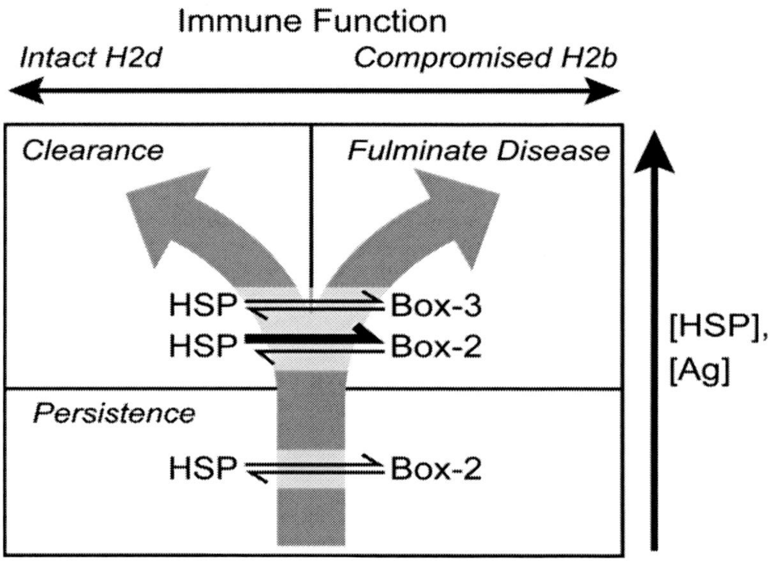

Figure 11. A model for the biological consequences of MeV NTAIL interaction with hsp72. Low levels of hsp72 reflect constitutive cellular expression or virus-induced hsp72, where the concentration supports high affinity interactions with Box-2 of NTAIL that are sufficient to support a persistent mode of replication. Levels of hsp72 can be increased by physiological stimuli such as fever, resulting in increased hsp72 interactions with Box-2 and also allowing for significant low affinity interactions with Box-3 that support enhanced viral transcription. Within an immunocompetent host, the resultant increased viral antigenic burden can facilitate adaptive immune responses leading to viral clearance. This is modeled in brain infection of mice of the H-2d haplotype. Within an immune-incompetent host, the increased viral gene expression can lead to enhanced CPE and thus virus-induced morbidity and mortality. This is modeled in brain infection of mice of the H-2b haplotype. Linkage between increased hsp72/Box-2 interaction, viral genome replication, and the impact upon the outcome of in vivo infection remains to be shown.

In this context, host immune responses both contain and benefit from the hsp72-dependent burst in viral antigen production during the acute phase of infection. This relationship would indicate a protective role for fever-induced hsp72, and would further suggest an immune selectional pressure for the emergence of viral N_{TAIL} variants that are hypo-responsive to hsp72 (e.g., N-522D). Prevalence of the MeV N_{TAIL} 522D sequence variant supports this possibility, particularly since the N522D substitution diminishes viral fitness [24]. Conversely, elevated hsp72 levels would be detrimental to the host if infected with an hsp72 responsive virus and adaptive immune responses were not adequate to contain the hsp72-dependent increase in viral gene expression. The result would be enhanced neurovirulence and more fulminate clinical disease. A 522D Box-3 variant would be predisposed to persistence, regardless of hsp72 levels or immune status of the host. All of these scenarios define hsp72 and N_{TAIL} as significant determinants of viral virulence, with immunological status determining the host beneficial versus detrimental function of hsp72.

It is likely that hsp72 would have similarly disparate effects upon MeV pathogenicity outside of the CNS as within, with outcome being dependent upon host immune status. MeV exhibits an hsp72 responsiveness that is similar between continuous cell lines of epithelial/mesenchymal origin (Vero, HEp-2, 293) [116, 150, 162, 163] and those of nervous system origin (human astrocytoma, murine neuroblastoma) [24, 151]. Additional variables that must be considered are potential hsp72-dependent effects of MV on antiviral immunity and the effect that increases in viral transcription and/or genome replication may have on infectious viral progeny release. Ultimately, it is the impact of hsp72 on peripheral infectious viral progeny release that will define the degree to which MV has exploited tissue elevations in hsp72 levels in support of viral transmission.

Acknowledgements

This work was supported by funds from the National Institute of Neurological Disorders and Stroke (R01NS31693).

References

[1] Agostini, I., S. Popov, J. Li, L. Dubrovsky, T. Hao, and M. Bukrinsky. 2000. Heat-shock protein 70 can replace viral protein R of HIV-1 during nuclear import of the viral preintegration complex. *Exp. Cell Res.* 259:398-403.

[2] Alfano, C. and R. McMacken. 1989. Heat shock protein-mediated disassembly of nucleoprotein structures is required for the initiation of bacteriophage lambda DNA replication. *J. Biol. Chem.* 264:10709-10718.

[3] Andrews, J. M., G. C. Newbound, M. Oglesbee, J. N. Brady, and M. D. Lairmore. 1997. The cellular stress response enhances human T-cell lymphotropic virus type 1 basal gene expression through the core promoter region of the long terminal repeat. *J. Virol.* 71:741-745.

[4] Asea, A. 2005. Stress proteins and initiation of immune response: chaperokine activity of hsp72. *Exerc. Immunol. Rev.* 11:34-45.

[5] Baler, R., J. Zou, and R. Voellmy. 1996. Evidence for a role of Hsp70 in the regulation of the heat shock response in mammalian cells. *Cell Stress*. Chaperones. 1:33-39.

[6] Barouch, W., K. Prasad, L. E. Greene, and E. Eisenberg. 1994. ATPase activity associated with the uncoating of clathrin baskets by Hsp70. *J. Biol. Chem*. 269:28563-28568.

[7] Beck, J. and M. Nassal. 2003. Efficient Hsp90-independent in vitro activation by Hsc70 and Hsp40 of duck hepatitis B virus reverse transcriptase, an assumed Hsp90 client protein. *J. Biol. Chem*. 278:36128-36138.

[8] Bhella, D., A. Ralph, L. B. Murphy, and R. P. Yeo. 2002. Significant differences in nucleocapsid morphology within the Paramyxoviridae. *J. Gen. Virol*. 83:1831-1839.

[9] Bhella, D., A. Ralph, and R. P. Yeo. 2004. Conformational flexibility in recombinant measles virus nucleocapsids visualised by cryo-negative stain electron microscopy and real-space helical reconstruction. *J. Mol. Biol.* 340:319-331.

[10] Blixenkrone-Moller, M., A. Bernard, A. Bencsik, N. Sixt, L. E. Diamond, J. S. Logan, and T. F. Wild. 1998. Role of CD46 in measles virus infection in CD46 transgenic mice. *Virology* 249:238-248.

[11] Bolt, G. 2001. The measles virus (MV) glycoproteins interact with cellular chaperones in the endoplasmic reticulum and MV infection upregulates chaperone expression. *Arch. Virol.* 146:2055-2068.

[12] Bonnycastle, L. L., C. E. Yu, C. R. Hunt, B. J. Trask, K. P. Clancy, J. L. Weber, D. Patterson, and G. D. Schellenberg. 1994. Cloning, sequencing, and mapping of the human chromosome 14 heat shock protein gene (HSPA2). *Genomics* 23:85-93.

[13] Bourhis, J. M., B. Canard, and S. Longhi. 2006. Structural disorder within the replicative complex of measles virus: Functional implications. *Virology* 344:94-110.

[14] Bourhis, J. M., K. Johansson, V. Receveur-Brechot, C. J. Oldfield, K. A. Dunker, B. Canard, and S. Longhi. 2004. The C-terminal domain of measles virus nucleoprotein belongs to the class of intrinsically disordered proteins that fold upon binding to their physiological partner. *Virus Res.* 99:157-167.

[15] Bourhis, J. M., V. Receveur-Brechot, M. Oglesbee, X. Zhang, M. Buccellato, H. Darbon, B. Canard, S. Finet, and S. Longhi. 2005. The intrinsically disordered C-terminal domain of the measles virus nucleoprotein interacts with the C-terminal domain of the phosphoprotein via two distinct sites and remains predominantly unfolded. *Protein Sci.* 14:1975-1992.

[16] Brasier, A. R., H. Spratt, Z. Wu, I. Boldogh, Y. Zhang, R. P. Garofalo, A. Casola, J. Pashmi, A. Haag, B. Luxon, and A. Kurosky. 2004. Nuclear heat shock response and novel nuclear domain 10 reorganization in respiratory syncytial virus-infected a549 cells identified by high-resolution two-dimensional gel electrophoresis. *J. Virol.* 78:11461-11476.

[17] Brehmer, D., S. Rudiger, C. S. Gassler, D. Klostermeier, L. Packschies, J. Reinstein, M. P. Mayer, and B. Bukau. 2001. Tuning of chaperone activity of Hsp70 proteins by modulation of nucleotide exchange. *Nat. Struct. Biol.* 8:427-432.

[18] Brooks, G. F., J. S. Butel, and S. A. Morse. 1998. Paramyxovirus and rubella virus, p. 507-527. *In* J. P. Butler, J. Ransom, and E. Ryan (eds.), *Adelberg's Microbiology.* Appleton and Lange, Stamford, Conn.

[19] Brown, G., H. W. Rixon, J. Steel, T. P. McDonald, A. R. Pitt, S. Graham, and R. J. Sugrue. 2005. Evidence for an association between heat shock protein 70 and the respiratory syncytial virus polymerase complex within lipid-raft membranes during virus infection. *Virology* 338:69-80.

[20] Campbell, K. S., K. P. Mullane, I. A. Aksoy, H. Stubdal, J. Zalvide, J. M. Pipas, P. A. Silver, T. M. Roberts, B. S. Schaffhausen, and J. A. DeCaprio. 1997. DnaJ/hsp40 chaperone domain of SV40 large T antigen promotes efficient viral DNA replication. *Genes Dev.* 11:1098-1110.

[21] Caracciolo, V., K. Reiss, K. Khalili, F. G. De, and A. Giordano. 2006. Role of the interaction between large T antigen and Rb family members in the oncogenicity of JC virus. *Oncogene* 25:5294-5301.

[22] Carsillo, T., M. Carsillo, S. Niewiesk, D. Vasconcelos, and M. Oglesbee. 2004. Hyperthermic pre-conditioning promotes measles virus clearance from brain in a mouse model of persistent infection. *Brain Res.* 1004:73-82.

[23] Carsillo, T., Z. Traylor, C. Choi, S. Niewiesk, and M. Oglesbee. 2006. Hsp72 - A Host Determinant of Measles Virus Neurovirulence. *J. Virol.*

[24] Carsillo, T., X. Zhang, D. Vasconcelos, S. Niewiesk, and M. Oglesbee. 2006. A single codon in the nucleocapsid protein C terminus contributes to in vitro and in vivo fitness of Edmonston measles virus. *J. Virol.* 80:2904-2912.

[25] Carsillo, T., X. Zhang, D. Vasconcelos, S. Niewiesk, and M. Oglesbee. *J.Virol.*, in press.

[26] Chan, C. P., K. L. Siu, K. T. Chin, K. Y. Yuen, B. Zheng, and D. Y. Jin. 2006. Modulation of the unfolded protein response by the severe acute respiratory syndrome coronavirus spike protein. *J. Virol.* 80:9279-9287.

[27] Chen, W., Y. Lin, C. Liao, and S. Hsieh. 2000. Modulatory effects of the human heat shock protein 70 on DNA vaccination. *J. Biomed. Sci.* 7:412-419.

[28] Cho, D. Y., G. H. Yang, C. J. Ryu, and H. J. Hong. 2003. Molecular chaperone GRP78/BiP interacts with the large surface protein of hepatitis B virus in vitro and in vivo. *J. Virol.* 77:2784-2788.

[29] Chromy, L. R., J. M. Pipas, and R. L. Garcea. 2003. Chaperone-mediated in vitro assembly of Polyomavirus capsids. *Proc. Natl. Acad. Sci. USA* 100:10477-10482.

[30] Collins, P. L. and L. E. Hightower. 1982. Newcastle disease virus stimulates the cellular accumulation of stress (heat shock) mRNAs and proteins. *J. Virol.* 44:703-707.

[31] Connell, P., C. A. Ballinger, J. Jiang, Y. Wu, L. J. Thompson, J. Hohfeld, and C. Patterson. 2001. The co-chaperone CHIP regulates protein triage decisions mediated by heat-shock proteins. *Nat. Cell Biol.* 3:93-96.

[32] Cripe, T. P., S. E. Delos, P. A. Estes, and R. L. Garcea. 1995. In vivo and in vitro association of hsc70 with polyomavirus capsid proteins. *J. Virol.* 69:7807-7813.

[33] Cutts, F. T., R. H. Henderson, C. J. Clements, R. T. Chen, and P. A. Patriarca. 1991. Principles of measles control. *Bull. World Health Organ* 69:1-7.

[34] Dai, Q., C. Zhang, Y. Wu, H. McDonough, R. A. Whaley, V. Godfrey, H. H. Li, N. Madamanchi, W. Xu, L. Neckers, D. Cyr, and C. Patterson. 2003. CHIP activates HSF1 and confers protection against apoptosis and cellular stress. *EMBO J.* 22:5446-5458.

[35] Damania, B., R. Mital, and J. C. Alwine. 1998. Simian virus 40 large T antigen interacts with human TFIIB-related factor and small nuclear RNA-activating protein complex for transcriptional activation of TATA-containing polymerase III promoters. *Mol. Cell Biol.* 18:1331-1338.

[36] Diehl, K. A., E. Crawford, P. D. Shinko, R. D. Tallman, Jr., and M. J. Oglesbee. 2000. Alterations in hemostasis associated with hyperthermia in a canine model. *Am. J. Hematol.* 64:262-270.

[37] Dragovic, Z., S. A. Broadley, Y. Shomura, A. Bracher, and F. U. Hartl. 2006. Molecular chaperones of the Hsp110 family act as nucleotide exchange factors of Hsp70s. *EMBO J.* 25:2519-2528.

[38] Duprex, W. P., I. Duffy, S. McQuaid, L. Hamill, S. L. Cosby, M. A. Billeter, J. Schneider-Schaulies, V. ter Meulen, and B. K. Rima. 1999. The H gene of rodent brain-adapted measles virus confers neurovirulence to the Edmonston vaccine strain. *J. Virol.* 73:6916-6922.

[39] Esiri, M. M. and P. G. Kennedy. 1997. Viral Diseases, p. 3-63. *In* D. I. Graham and P. L. Lantos (eds.), *Greenfield's Neuropathology*. Arnold, New York.

[40] Esiri, M. M., D. R. Oppenheimer, B. Brownell, and M. Haire. 1982. Distribution of measles antigen and immunoglobulin-containing cells in the CNS in subacute sclerosing panencephalitis (SSPE) and atypical measles encephalitis. *J. Neurol. Sci.* 53:29-43.

[41] Finke, D. and U. G. Liebert. 1994. CD4+ T cells are essential in overcoming experimental murine measles encephalitis. *Immunology* 83:184-189.

[42] Fourie, A. M., T. R. Hupp, D. P. Lane, B. C. Sang, M. S. Barbosa, J. F. Sambrook, and M. J. Gething. 1997. HSP70 binding sites in the tumor suppressor protein p53. *J. Biol. Chem.* 272:19471-19479.

[43] Fourie, A. M., J. F. Sambrook, and M. J. Gething. 1994. Common and divergent peptide binding specificities of hsp70 molecular chaperones. *J. Biol. Chem.* 269:30470-30478.

[44] Freeman, A. F., D. A. Jacobsohn, S. T. Shulman, W. J. Bellini, P. Jaggi, L. G. de, G. F. Keating, F. Kim, L. M. Pachman, M. Kletzel, and R. E. Duerst. 2004. A new complication of stem cell transplantation: measles inclusion body encephalitis. *Pediatrics* 114:e657-e660.

[45] Furlini, G., M. Vignoli, M. C. Re, D. Gibellini, E. Ramazzotti, G. Zauli, and P. M. La. 1994. Human immunodeficiency virus type 1 interaction with the membrane of CD4+ cells induces the synthesis and nuclear translocation of 70K heat shock protein. *J. Gen. Virol.* 75 (Pt 1):193-199.

[46] Gamer, J., G. Multhaup, T. Tomoyasu, J. S. McCarty, S. Rudiger, H. J. Schonfeld, C. Schirra, H. Bujard, and B. Bukau. 1996. A cycle of binding and release of the DnaK, DnaJ and GrpE chaperones regulates activity of the Escherichia coli heat shock transcription factor sigma32. *EMBO J.* 15:607-617.

[47] Garimella, R., X. Liu, W. Qiao, X. Liang, E. R. Zuiderweg, M. I. Riley, and S.R. Van Doren. 2006. Hsc70 contacts helix III of the J domain from polyomavirus T antigens: addressing a dilemma in the chaperone hypothesis of how they release E2F from pRb. *Biochemistry* 45:6917-6929.

[48] Gassler, C. S., T. Wiederkehr, D. Brehmer, B. Bukau, and M. P. Mayer. 2001. Bag-1M accelerates nucleotide release for human Hsc70 and Hsp70 and can act concentration-dependent as positive and negative cofactor. *J. Biol. Chem.* 276:32538-32544.

[49] Gaudin, Y. 1997. Folding of rabies virus glycoprotein: epitope acquisition and interaction with endoplasmic reticulum chaperones. *J. Virol.* 71:3742-3750.

[50] Gerner, E. W., E. M. Hersh, M. Pennington, T. C. Tsang, D. Harris, F. Vasanwala, and J. Brailey. 2000. Heat-inducible vectors for use in gene therapy. *Int. J. Hyperthermia* 16:171-181.

[51] Gething, M. J., S. Blond-Elguindi, J. Buchner, A. Fourie, G. Knarr, S. Modrow, L. Nanu, M. Segal, and J. Sambrook. 1995. Binding sites for Hsp70 molecular chaperones in natural proteins. Cold Spring Harb. *Symp. Quant. Biol.* 60:417-428.

[52] Giffard, R. G., L. Xu, H. Zhao, W. Carrico, Y. Ouyang, Y. Qiao, R. Sapolsky, G. Steinberg, B. Hu, and M. A. Yenari. 2004. Chaperones, protein aggregation, and brain protection from hypoxic/ischemic injury. *J. Exp. Biol.* 207:3213-3220.

[53] Glotzer, J. B., M. Saltik, S. Chiocca, A. I. Michou, P. Moseley, and M. Cotten. 2000. Activation of heat-shock response by an adenovirus is essential for virus replication. *Nature* 407:207-211.

[54] Good, R. A. and S. J. Zak. 1956. Disturbances in gamma globulin synthesis as experiments of nature. *Pediatrics* 18:109-149.

[55] Griffin, D. E., J. Mullinix, O. Narayan, and R. T. Johnson. 1974. Age dependence of viral expression: comparative pathogenesis of two rodent-adapted strains of measles virus in mice. *Infect. Immun.* 9:690-695.

[56] Gurer, C., A. Cimarelli, and J. Luban. 2002. Specific incorporation of heat shock protein 70 family members into primate lentiviral virions. *J. Virol.* 76:4666-4670.

[57] Halford, W. P., B. M. Gebhardt, and D. J. Carr. 1996. Mechanisms of herpes simplex virus type 1 reactivation. *J. Virol.* 70:5051-5060.

[58] Hammond, C. and A. Helenius. 1994. Folding of VSV G protein: sequential interaction with BiP and calnexin. *Science* 266:456-458.

[59] Hartl, F. U. 1996. Molecular chaperones in cellular protein folding. Nature 381:571-579.

[60] Hashimoto, K., M. Baba, K. Gohnai, M. Sato, and S. Shigeta. 1996. Heat shock induces HIV-1 replication in chronically infected promyelocyte cell line OM10.1. *Arch. Virol.* 141:439-447.

[61] Haviv, Y. S., J. L. Blackwell, H. Li, M. Wang, X. Lei, and D. T. Curiel. 2001. Heat shock and heat shock protein 70i enhance the oncolytic effect of replicative adenovirus. *Cancer Res.* 61:8361-8365.

[62] Heggeness, M. H., A. Scheid, and P. W. Choppin. 1980. Conformation of the helical nucleocapsids of paramyxoviruses and vesicular stomatitis virus: reversible coiling and uncoiling induced by changes in salt concentration. *Proc. Natl. Acad. Sci. USA* 77:2631-2635.

[63] Heimbach, J. K., L. L. Reznikov, C. M. Calkins, T. N. Robinson, C. A. Dinarello, A. H. Harken, and X. Meng. 2001. TNF receptor I is required for induction of macrophage heat shock protein 70. *Am. J. Physiol. Cell Physiol.* 281:C241-C247.

[64] Heller, M., D. Vasconcelos, J. Cummins, and M. Oglesbee. 1998. Interferon-alpha inhibits the emergence of cellular stress response-dependent morbillivirus large plaque variants. *Antiviral Res.* 38:195-207.

[65] Hennessy, F., W. S. Nicoll, R. Zimmermann, M. E. Cheetham, and G. L. Blatch. 2005. Not all J domains are created equal: implications for the specificity of Hsp40-Hsp70 interactions. *Protein Sci.* 14:1697-1709.

[66] Hernandez, M. P., W. P. Sullivan, and D. O. Toft. 2002. The assembly and intermolecular properties of the hsp70-Hop-hsp90 molecular chaperone complex. *J. Biol. Chem.* 277:38294-38304.

[67] Jacobson, S., R. P. Sekaly, C. L. Jacobson, H. F. McFarland, and E. O. Long. 1989. HLA class II-restricted presentation of cytoplasmic measles virus antigens to cytotoxic T cells. *J. Virol.* 63:1756-1762.

[68] Jiang, J., K. Prasad, E. M. Lafer, and R. Sousa. 2005. Structural basis of interdomain communication in the Hsc70 chaperone. *Mol. Cell* 20:513-524.

[69] Johansson, K., J. M. Bourhis, V. Campanacci, C. Cambillau, B. Canard, and S. Longhi. 2003. Crystal structure of the measles virus phosphoprotein domain responsible for the induced folding of the C-terminal domain of the nucleoprotein. *J. Biol. Chem.* 278:44567-44573.

[70] Kaghad, M., X. Dumont, P. Chalon, J. M. Lelias, N. Lamande, M. Lucas, M. Lazar, and D. Caput. 1990. Nucleotide sequences of cDNAs alpha and gamma enolase mRNAs from mouse brain. *Nucleic Acids Res.* 18:3638.

[71] Kanelakis, K. C., P. J. Murphy, M. D. Galigniana, Y. Morishima, S. Takayama, J. C. Reed, D. O. Toft, and W. B. Pratt. 2000. hsp70 interacting protein Hip does not affect glucocorticoid receptor folding by the hsp90-based chaperone machinery except to oppose the effect of BAG-1. *Biochemistry* 39:14314-14321.

[72] Karlin, D., S. Longhi, and B. Canard. 2002. Substitution of two residues in the measles virus nucleoprotein results in an impaired self-association. *Virology* 302:420-432.

[73] Katayama, Y., H. Hotta, A. Nishimura, Y. Tatsuno, and M. Homma. 1995. Detection of measles virus nucleoprotein mRNA in autopsied brain tissues. *J. Gen. Virol.* 76 (Pt 12):3201-3204.

[74] Kiang, J. G. and G. C. Tsokos. 1998. Heat shock protein 70 kDa: molecular biology, biochemistry, and physiology. *Pharmacol. Ther.* 80:183-201.

[75] Kobayashi, K., E. Ohgitani, Y. Tanaka, M. Kita, and J. Imanishi. 1994. Herpes simplex virus-induced expression of 70 kDa heat shock protein (HSP70) requires early protein synthesis but not viral DNA replication. *Microbiol. Immunol.* 38:321-325.

[76] Koestner, A. and J. F. Long. 1970. Ultrastructure of canine distemper virus in explant tissue cultures of canine cerebellum. *Lab. Invest.* 23:196-201.

[77] Kreis, S., E. Vardas, and T. Whistler. 1997. Sequence analysis of the nucleocapsid gene of measles virus isolates from South Africa identifies a new genotype. *J. Gen. Virol.* 78 (Pt 7):1581-1587.

[78] Laine, D., J. M. Bourhis, S. Longhi, M. Flacher, L. Cassard, B. Canard, C. Sautes-Fridman, C. Rabourdin-Combe, and H. Valentin. 2005. Measles virus nucleoprotein induces cell-proliferation arrest and apoptosis through NTAIL-NR and NCORE-FcgammaRIIB1 interactions, respectively. *J. Gen. Virol.* 86:1771-1784.

[79] Laine, D., M. C. Trescol-Biemont, S. Longhi, G. Libeau, J. C. Marie, P. O. Vidalain, O. Azocar, A. Diallo, B. Canard, C. Rabourdin-Combe, and H. Valentin. 2003. Measles virus (MV) nucleoprotein binds to a novel cell surface receptor distinct from FcgammaRII via its C-terminal domain: role in MV-induced immunosuppression. *J. Virol.* 77:11332-11346.

[80] Lamb, R. and D. Kolakofsky. 2005. Paramyxoviridae: the viruses and their replication, p. 1305-1340. *In* D. M. Knipe and P. M. Howley (eds.), *Fields Virology*. Lippincott, Williams and Wilkins, Philadelphia, PA.

[81] Lambert, C. and R. Prange. 2003. Chaperone action in the posttranslational topological reorientation of the hepatitis B virus large envelope protein: Implications for translocational regulation. *Proc. Natl. Acad. Sci. USA* 100:5199-5204.

[82] Laufen, T., M. P. Mayer, C. Beisel, D. Klostermeier, A. Mogk, J. Reinstein, and B. Bukau. 1999. Mechanism of regulation of hsp70 chaperones by DnaJ cochaperones. *Proc. Natl. Acad. Sci. USA* 96:5452-5457.

[83] Lawrence, D. M., C. E. Patterson, T. L. Gales, J. L. D'Orazio, M. M. Vaughn, and G. F. Rall. 2000. Measles virus spread between neurons requires cell contact but not CD46 expression, syncytium formation, or extracellular virus production. *J. Virol.* 74:1908-1918.

[84] Lawrence, D. M., M. M. Vaughn, A. R. Belman, J. S. Cole, and G. F. Rall. 1999. Immune response-mediated protection of adult but not neonatal mice from neuron-restricted measles virus infection and central nervous system disease. *J. Virol.* 73:1795-1801.

[85] Leone, G., M. C. Coffey, R. Gilmore, R. Duncan, L. Maybaum, and P. W. Lee. 1996. C-terminal trimerization, but not N-terminal trimerization, of the reovirus cell attachment protein Is a posttranslational and Hsp70/ATP-dependent process. *J. Biol. Chem.* 271:8466-8471.

[86] Li, J. and B. Sha. 2005. Structure-based mutagenesis studies of the peptide substrate binding fragment of type I heat-shock protein 40. *Biochem. J.* 386:453-460.

[87] Liebert, U. G. and D. Finke. 1995. Measles virus infections in rodents. *Curr. Top. Microbiol. Immunol.* 191:149-166.

[88] Liu, J. S., S. R. Kuo, A. M. Makhov, D. M. Cyr, J. D. Griffith, T. R. Broker, and L. T. Chow. 1998. Human Hsp70 and Hsp40 chaperone proteins facilitate human papillomavirus-11 E1 protein binding to the origin and stimulate cell-free DNA replication. *J. Biol. Chem.* 273:30704-30712.

[89] Longhi, S., V. Receveur-Brechot, D. Karlin, K. Johansson, H. Darbon, D. Bhella, R. Yeo, S. Finet, and B. Canard. 2003. The C-terminal domain of the measles virus nucleoprotein is intrinsically disordered and folds upon binding to the C-terminal moiety of the phosphoprotein. *J. Biol. Chem.* 278:18638-18648.

[90] Macejak, D. G. and R. B. Luftig. 1991. Association of HSP70 with the adenovirus type 5 fiber protein in infected HEp-2 cells. *Virology* 180:120-125.

[91] Macejak, D. G. and P. Sarnow. 1992. Association of heat shock protein 70 with enterovirus capsid precursor P1 in infected human cells. *J. Virol.* 66:1520-1527.

[92] Mayer, M. P. 2005. Recruitment of Hsp70 chaperones: a crucial part of viral survival strategies. *Rev. Physiol Biochem. Pharmacol.* 153:1-46.

[93] Milarski, K. L. and R. I. Morimoto. 1986. Expression of human HSP70 during the synthetic phase of the cell cycle. *Proc. Natl. Acad. Sci. USA* 83:9517-9521.

[94] Milarski, K. L. and R. I. Morimoto. 1989. Mutational analysis of the human HSP70 protein: distinct domains for nucleolar localization and adenosine triphosphate binding. *J. Cell Biol.* 109:1947-1962.

[95] Minami, Y., J. Hohfeld, K. Ohtsuka, and F. U. Hartl. 1996. Regulation of the heat-shock protein 70 reaction cycle by the mammalian DnaJ homolog, Hsp40. *J. Biol. Chem.* 271:19617-19624.

[96] Misselwitz, B., O. Staeck, and T. A. Rapoport. 1998. J proteins catalytically activate Hsp70 molecules to trap a wide range of peptide sequences. *Mol. Cell* 2:593-603.

[97] Mizzen, L. A. and W. J. Welch. 1988. Characterization of the thermotolerant cell. I. Effects on protein synthesis activity and the regulation of heat-shock protein 70 expression. *J. Cell Biol.* 106:1105-1116.

[98] Morrison-Bogorad, M., A. L. Zimmerman, and S. Pardue. 1995. Heat-shock 70 messenger RNA levels in human brain: correlation with agonal fever. *J. Neurochem.* 64:235-246.

[99] Mrkic, B., J. Pavlovic, T. Rulicke, P. Volpe, C. J. Buchholz, D. Hourcade, J. P. Atkinson, A. Aguzzi, and R. Cattaneo. 1998. Measles virus spread and pathogenesis in genetically modified mice. *J. Virol.* 72:7420-7427.

[100] Murakami, T., H. Ohmori, S. Gotoh, T. Tsuda, R. Ohya, S. Akiya, and K. Higashi. 1991. Down modulation of N-myc, heat-shock protein 70, and nucleolin during the differentiation of human neuroblastoma cells. *J. Biochem.* (Tokyo) 110:146-150.

[101] Neighbour, P. A., B. Rager-Zisman, and B. R. Bloom. 1978. Susceptibility of mice to acute and persistent measles infection. *Infect. Immun.* 21:764-770.

[102] Neumeister, C. and S. Niewiesk. 1998. Recognition of measles virus-infected cells by CD8+ T cells depends on the H-2 molecule. *J. Gen. Virol.* 79 (Pt 11):2583-2591.

[103] Niewiesk, S., U. Brinckmann, B. Bankamp, S. Sirak, U. G. Liebert, and V. ter Meulen. 1993. Susceptibility to measles virus-induced encephalitis in mice correlates with impaired antigen presentation to cytotoxic T lymphocytes. *J. Virol.* 67:75-81.

[104] Oglesbee, M. 1992. Intranuclear inclusions in paramyxovirus-induced encephalitis: evidence for altered nuclear body differentiation. *Acta Neuropathol.* (Berl) 84:407-415.

[105] Oglesbee, M. and S. Krakowka. 1993. Cellular stress response induces selective intranuclear trafficking and accumulation of morbillivirus major core protein. *Lab. Invest.* 68:109-117.

[106] Oglesbee, M., S. Ringler, and S. Krakowka. 1990. Interaction of canine distemper virus nucleocapsid variants with 70K heat-shock proteins. *J. Gen. Virol.* 71 (Pt 7):1585-1590.

[107] Oglesbee, M., L. Tatalick, J. Rice, and S. Krakowka. 1989. Isolation and characterization of canine distemper virus nucleocapsid variants. *J. Gen. Virol.* 70 (Pt 9):2409-2419.

[108] Oglesbee, M. J., S. Alldinger, D. Vasconcelos, K. A. Diehl, P. D. Shinko, W. Baumgartner, R. Tallman, and M. Podell. 2002. Intrinsic thermal resistance of the canine brain. *Neuroscience* 113:55-64.

[109] Oglesbee, M. J., H. Kenney, T. Kenney, and S. Krakowka. 1993. Enhanced production of morbillivirus gene-specific RNAs following induction of the cellular stress response in stable persistent infection. *Virology* 192:556-567.

[110] Oglesbee, M. J., Z. Liu, H. Kenney, and C. L. Brooks. 1996. The highly inducible member of the 70 kDa family of heat shock proteins increases canine distemper virus polymerase activity. *J. Gen. Virol.* 77 (Pt 9):2125-2135.

[111] Oglesbee, M. J., M. Pratt, and T. Carsillo. 2002. Role for heat shock proteins in the immune response to measles virus infection. *Viral Immunol.* 15:399-416.

[112] Ohgitani, E., K. Kobayashi, K. Takeshita, and J. Imanishi. 1998. Induced expression and localization to nuclear-inclusion bodies of hsp70 in varicella-zoster virus-infected human diploid fibroblasts. *Microbiol. Immunol.* 42:755-760.

[113] Otteken, A. and B. Moss. 1996. Calreticulin interacts with newly synthesized human immunodeficiency virus type 1 envelope glycoprotein, suggesting a chaperone function similar to that of calnexin. *J. Biol. Chem.* 271:97-103.

[114] Pack, C. D., U. Kumaraguru, S. Suvas, and B. T. Rouse. 2005. Heat-shock protein 70 acts as an effective adjuvant in neonatal mice and confers protection against challenge with Herpes Simplex Virus. *Vaccine* 23:3526-3534.

[115] Pardue, S., S. Wang, M. M. Miller, and M. Morrison-Bogorad. 2006. Elevated levels of inducible heat shock 70 proteins in human brain. *Neurobiol. Aging.*

[116] Parks, C. L., R. A. Lerch, P. Walpita, M. S. Sidhu, and S. A. Udem. 1999. Enhanced measles virus cDNA rescue and gene expression after heat shock. *J. Virol.* 73:3560-3566.

[117] Peluso, R. W., R. A. Lamb, and P. W. Choppin. 1978. Infection with paramyxoviruses stimulates synthesis of cellular polypeptides that are also stimulated in cells transformed by Rous sarcoma virus or deprived of glucose. *Proc. Natl. Acad. Sci. USA* 75:6120-6124.

[118] Perez-Vargas, J., P. Romero, S. Lopez, and C. F. Arias. 2006. The peptide-binding and ATPase domains of recombinant hsc70 are required to interact with rotavirus and reduce its infectivity. *J. Virol.* 80:3322-3331.

[119] Peterson, J. R., A. Ora, P. N. Van, and A. Helenius. 1995. Transient, lectin-like association of calreticulin with folding intermediates of cellular and viral glycoproteins. *Mol. Biol. Cell* 6:1173-1184.

[120] Phillips, B., K. Abravaya, and R. I. Morimoto. 1991. Analysis of the specificity and mechanism of transcriptional activation of the human hsp70 gene during infection by DNA viruses. *J. Virol.* 65:5680-5692.

[121] Pilon, M. and R. Schekman. 1999. Protein translocation: how Hsp70 pulls it off. *Cell* 97:679-682.

[122] Qanungo, K. R., D. Shaji, M. Mathur, and A. K. Banerjee. 2004. Two RNA polymerase complexes from vesicular stomatitis virus-infected cells that carry out transcription and replication of genome RNA. *Proc. Natl. Acad. Sci. USA* 101:5952-5957.

[123] Rall, G. F., M. Manchester, L. R. Daniels, E. M. Callahan, A. R. Belman, and M. B. Oldstone. 1997. A transgenic mouse model for measles virus infection of the brain. *Proc. Natl. Acad. Sci. USA* 94:4659-4663.

[124] Ravagnan, L., S. Gurbuxani, S. A. Susin, C. Maisse, E. Daugas, N. Zamzami, T. Mak, M. Jaattela, J. M. Penninger, C. Garrido, and G. Kroemer. 2001. Heat-shock protein 70 antagonizes apoptosis-inducing factor. *Nat. Cell Biol.* 3:839-843.

[125] Robbins, S. J., R. H. Bussell, and F. Rapp. 1980. Isolation and partial characterization of two forms of cytoplasmic nucleocapsids from measles virus-infected cells. *J. Gen. Virol.* 47:301-310.

[126] Roux, L. 1990. Selective and transient association of Sendai virus HN glycoprotein with BiP. *Virology* 175:161-166.

[127] Sagara, Y., C. Ishida, Y. Inoue, H. Shiraki, and Y. Maeda. 1998. 71-kilodalton heat shock cognate protein acts as a cellular receptor for syncytium formation induced by human T-cell lymphotropic virus type 1. *J. Virol.* 72:535-541.

[128] Sakae, H., M. Kohase, T. Kurata, and H. Yoshikura. 1997. Isolation of a measles virus variant: protection of newborn mice from measles encephalitis by 24 h prior intracerebral inoculation with the variant. *Arch. Virol.* 142:1937-1952.

[129] Saphire, A. C., T. Guan, E. C. Schirmer, G. R. Nemerow, and L. Gerace. 2000. Nuclear import of adenovirus DNA in vitro involves the nuclear protein import pathway and hsc70. *J. Biol. Chem.* 275:4298-4304.

[130] Scheufler, C., A. Brinker, G. Bourenkov, S. Pegoraro, L. Moroder, H. Bartunik, F. U. Hartl, and I. Moarefi. 2000. Structure of TPR domain-peptide complexes: critical elements in the assembly of the Hsp70-Hsp90 multichaperone machine. *Cell* 101:199-210.

[131] Schneikert, J., S. Hubner, G. Langer, T. Petri, M. Jaattela, J. Reed, and A. C. Cato. 2000. Hsp70-RAP46 interaction in downregulation of DNA binding by glucocorticoid receptor. *EMBO J.* 19:6508-6516.

[132] Schroder, M. and R. J. Kaufman. 2005. The mammalian unfolded protein response. *Annu. Rev. Biochem.* 74:739-789.

[133] Schumacher, R. J., R. Hurst, W. P. Sullivan, N. J. McMahon, D. O. Toft, and R. L. Matts. 1994. ATP-dependent chaperoning activity of reticulocyte lysate. *J. Biol. Chem.* 269:9493-9499.

[134] Shomura, Y., Z. Dragovic, H. C. Chang, N. Tzvetkov, J. C. Young, J. L. Brodsky, V. Guerriero, F. U. Hartl, and A. Bracher. 2005. Regulation of Hsp70 function by HspBP1: structural analysis reveals an alternate mechanism for Hsp70 nucleotide exchange. *Mol. Cell* 17:367-379.

[135] Snider, B. J., D. Lobner, K. A. Yamada, and D. W. Choi. 1998. Conditioning heat stress reduces excitotoxic and apoptotic components of oxygen-glucose deprivation-induced neuronal death in vitro. *J. Neurochem.* 70:120-129.

[136] Song, J., M. Takeda, and R. I. Morimoto. 2001. Bag1-Hsp70 mediates a physiological stress signalling pathway that regulates Raf-1/ERK and cell growth. *Nat. Cell Biol.* 3:276-282.

[137] Sousa, M. C. and D. B. McKay. 1998. The hydroxyl of threonine 13 of the bovine 70-kDa heat shock cognate protein is essential for transducing the ATP-induced conformational change. *Biochemistry* 37:15392-15399.

[138] Strandell, E., K. Buschard, J. Saldeen, and N. Welsh. 1995. Interleukin-1 beta induces the expression of hsp70, heme oxygenase and Mn-SOD in FACS-purified rat islet beta-cells, but not in alpha-cells. *Immunol. Lett.* 48:145-148.

[139] Su, C. Y., K. Y. Chong, J. Chen, S. Ryter, R. Khardori, and C. C. Lai. 1999. A physiologically relevant hyperthermia selectively activates constitutive hsp70 in H9c2 cardiac myoblasts and confers oxidative protection. *J. Mol. Cell Cardiol.* 31:845-855.

[140] Takahashi, N., R. Sasaki, J. Takahashi, S. Takayama, J. C. Reed, and T. Andoh. 2001. BAG-1M, an isoform of Bcl-2-interacting protein BAG-1, enhances gene expression driven by CMV promoter. *Biochem. Biophys. Res. Commun.* 286:807-814.

[141] Takayama, S., D. N. Bimston, S. Matsuzawa, B. C. Freeman, C. Aime-Sempe, Z. Xie, R. I. Morimoto, and J. C. Reed. 1997. BAG-1 modulates the chaperone activity of Hsp70/Hsc70. *EMBO J.* 16:4887-4896.

[142] Takayama, S. and J. C. Reed. 2001. Molecular chaperone targeting and regulation by BAG family proteins. *Nat. Cell Biol.* 3:E237-E241.

[143] Takayama, S., Z. Xie, and J. C. Reed. 1999. An evolutionarily conserved family of Hsp70/Hsc70 molecular chaperone regulators. *J. Biol. Chem.* 274:781-786.

[144] Tanguy Le, G. N. and P. E. Boehmer. 2002. Activation of the herpes simplex virus type-1 origin-binding protein (UL9) by heat shock proteins. *J. Biol. Chem.* 277:5660-5666.

[145] Tapparel, C., D. Maurice, and L. Roux. 1998. The activity of Sendai virus genomic and antigenomic promoters requires a second element past the leader template regions: a motif (GNNNNN)3 is essential for replication. *J. Virol.* 72:3117-3128.

[146] Theodorakis, N. G., D. Drujan, and M. A. De. 1999. Thermotolerant cells show an attenuated expression of Hsp70 after heat shock. *J. Biol. Chem.* 274:12081-12086.

[147] Theyssen, H., H. P. Schuster, L. Packschies, B. Bukau, and J. Reinstein. 1996. The second step of ATP binding to DnaK induces peptide release. *J. Mol. Biol.* 263:657-670.

[148] Uytdehaag, F. G., R. S. van Binnendijk, M. J. Kenter, and A. D. Osterhaus. 1994. Cytotoxic T lymphocyte responses against measles virus. *Curr. Top. Microbiol. Immunol.* 189:151-167.

[149] van Binnendijk, R. S., M. C. Poelen, K. C. Kuijpers, A. D. Osterhaus, and F. G. Uytdehaag. 1990. The predominance of CD8+ T cells after infection with measles virus suggests a role for CD8+ class I MHC-restricted cytotoxic T lymphocytes (CTL) in recovery from measles. Clonal analyses of human CD8+ class I MHC-restricted CTL. *J. Immunol.* 144:2394-2399.

[150] Vasconcelos, D., E. Norrby, and M. Oglesbee. 1998. The cellular stress response increases measles virus-induced cytopathic effect. *J. Gen. Virol.* 79 (Pt 7):1769-1773.

[151] Vasconcelos, D. Y., X. H. Cai, and M. J. Oglesbee. 1998. Constitutive overexpression of the major inducible 70 kDa heat shock protein mediates large plaque formation by measles virus. *J. Gen. Virol.* 79 (Pt 9):2239-2247.

[152] von Messling, V., V, G. Zimmer, G. Herrler, L. Haas, and R. Cattaneo. 2001. The hemagglutinin of canine distemper virus determines tropism and cytopathogenicity. *J. Virol.* 75:6418-6427.

[153] Wainberg, Z., M. Oliveira, S. Lerner, Y. Tao, and B. G. Brenner. 1997. Modulation of stress protein (hsp27 and hsp70) expression in CD4+ lymphocytic cells following acute infection with human immunodeficiency virus type-1. *Virology* 233:364-373.

[154] Weidinger, G., S. Czub, C. Neumeister, P. Harriott, V. ter Meulen, and S. Niewiesk. 2000. Role of CD4(+) and CD8(+) T cells in the prevention of measles virus-induced encephalitis in mice. *J. Gen. Virol.* 81:2707-2713.

[155] Weidinger, G., M. Ohlmann, B. Schlereth, G. Sutter, and S. Niewiesk. 2001. Vaccination with recombinant modified vaccinia virus Ankara protects against measles virus infection in the mouse and cotton rat model. *Vaccine* 19:2764-2768.

[156] Xiao, J., L. S. Kim, and T. R. Graham. 2006. Dissection of Swa2p/auxilin domain requirements for cochaperoning Hsp70 clathrin-uncoating activity in vivo. *Mol. Biol. Cell* 17:3281-3290.

[157] Xu, A., A. R. Bellamy, and J. A. Taylor. 1998. BiP (GRP78) and endoplasmin (GRP94) are induced following rotavirus infection and bind transiently to an endoplasmic reticulum-localized virion component. *J. Virol.* 72:9865-9872.

[158] Xu, C., B. Bailly-Maitre, and J. C. Reed. 2005. Endoplasmic reticulum stress: cell life and death decisions. *J. Clin. Invest* 115:2656-2664.

[159] Yager, J. Y. and J. Asselin. 1999. The effect of pre hypoxic-ischemic (HI) hypo and hyperthermia on brain damage in the immature rat. *Brain Res. Dev. Brain Res.* 117:139-143.

[160] Yoshida, H., K. Haze, H. Yanagi, T. Yura, and K. Mori. 1998. Identification of the cis-acting endoplasmic reticulum stress response element responsible for transcriptional induction of mammalian glucose-regulated proteins. Involvement of basic leucine zipper transcription factors. *J. Biol. Chem.* 273:33741-33749.

[161] Zhang, X., S. L. Baader, F. Bian, W. Muller, and J. Oberdick. 2001. High level Purkinje cell specific expression of green fluorescent protein in transgenic mice. Histochem. *Cell Biol.* 115:455-464.

[162] Zhang, X., J. M. Bourhis, S. Longhi, T. Carsillo, M. Buccellato, B. Morin, B. Canard, and M. Oglesbee. 2005. Hsp72 recognizes a P binding motif in the measles virus N protein C-terminus. *Virology* 337:162-174.

[163] Zhang, X., C. Glendening, H. Linke, C. L. Parks, C. Brooks, S. A. Udem, and M. Oglesbee. 2002. Identification and characterization of a regulatory domain on the carboxyl terminus of the measles virus nucleocapsid protein. *J. Virol.* 76:8737-8746.

[164] Zhang, X. and M. Oglesbee. 2003. Use of surface plasmon resonance for the measurement of low affinity binding interactions between HSP72 and measles virus nucleocapsid protein. *Biol. Proced.* Online. 5:170-181.

[165] Zhu, X., X. Zhao, W. F. Burkholder, A. Gragerov, C. M. Ogata, M. E. Gottesman, and W. A. Hendrickson. 1996. Structural analysis of substrate binding by the molecular chaperone DnaK. *Science* 272:1606-1614.

[166] Zylicz, M., F. W. King, and A. Wawrzynow. 2001. Hsp70 interactions with the p53 tumour suppressor protein. *EMBO J.* 20:4634-4638.

In: Measles Virus Nucleoprotein
Editors: S. Longhi, pp. 99-112
ISBN: 978-1-60021-629-9

Chapter IV

Interferon Regulatory Factor 3 as Cellular Partner of Measles Virus Nucleoprotein

Florence Herschke and Denis Gerlier*
Virologie et Pathogénèse Virale, Université Lyon 1,
CNRS, F-69372 Lyon, France

Abstract

IFN response, which plays an important part of cellular and immune response to measles virus infection, strongly relies on the activation of IRF-3, a constitutive cytoplasmic transcription factor. The sensing of a "danger" signal by receptors induces the activation of TBK1, which phosphorylates, IRF-3. Then, IRF-3 homo- or heterodimerizes, translocates to the nucleus and, along with NF-κB and AP-1, activates the expression of IFN-α/β and proinflammatory cytokines. Measles virus displays another IRF-3 activation route: the nucleoprotein (N), via its N_{TAIL} region, binds to IRF-3 IAD domain and somehow recruits TBK-1 to phosporylate IRF-3. Interestingly, only a subset of IRF-3-dependent genes (not including IFN-β) is activated through this interaction. The underlying mechanism is yet to be unravelled.

List of Abbreviations

For chemokine abbreviations and alternative names, please refer to [4]

AP-1	Activator protein-1
ATF-2	Activating transcription factor

* Corresponding author: VPV, CNRS UMR5537 - Université de Lyon 1, Faculté Laennec, F-69372 Lyon Cedex 08, France. E-mail : denis.gerlier@univ-lyon1.fr; Tel: (33) 4 78 77 86 18; Fax: (33) 4 78 77 87 54.

CBP	CREB binding protein
c-jun	Cellular-jun
CREB	cAMP response element-binding protein
CRM1	Chromosome maintenance 1
CyB	Cyclophilin B
DBD	DNA binding domain
DNA-PK	DNA protein kinase
d.p.i.	Days post-infection
dsRNA	Double stranded RNA
GRIP-1	Glucocorticoid receptor inhibitor protein-1
HMG-I (Y)	High mobility group-I (Y) protein
HSP-90	Heat shock protein-90
IAD	Interferon associated domain
IFN-α, β	Interferon-α, -β
IFNAR	IFN-α/β receptor
I-κB	Inhibitor of NF-κB
IKK-α, -β, -γ	I-κB kinase-α, -β, -γ
IPS-1	Interferon-β promoter stimulator-1 (also called MAVS/Cardif/VISA)
IRF-3, -7, -9	Interferon regulatory factor-3, -7, -9
ISG	Interferon stimulated gene
ISGF3	Interferon stimulated gene factor 3
ISRE	Interferon sensitive response element
JNK	Jun kinase
JUNK kinase	JNK kinase
KPNA3	Karyopherin-α3
L protein	(measles virus) large protein
MAP kinase	Mitogen-activated protein kinase
MDA-5	Melanoma differentiation associated gene-5
MeV	Measles virus
MKK3	MAP kinase kinase 3
N	(measles virus) nucleoprotein
NES	Nuclear export domain
NF-κB	Nuclear factor-κB
NLS	Nuclear localisation domain
P	(measles virus) phosphoprotein
PAMP	Pathogen associated molecular pattern
PCT	P C-terminal domain
Pin1	Peptidyl-prolyl isomerase 1
PMD	P multimerisation domain
PNT	P N-terminal domain
PRD-I, -II, -III, -IV	Positive regulatory domain-I, -II, -III, -IV
PRR	Pathogen recognition receptor
Qip1	A subtype of importin α

RIG-I	Retinoic acid-induced gene-I
RNA	Ribonucleic acid
SRR	Serine rich region
ssRNA	Single stranded RNA
STAT1, 2	Signal transducers and activators of transcription 1, 2
TBK-1	TANK binding kinase-1
TLR	Toll-like receptor
TRAF-6	TNF receptor associated factor-6

1. Introduction

IRF-3 belongs to the family of interferon regulatory factors (IRF) and acts a transactivator for the interferon-β (IFN-β) and various pro-inflammatory cytokine genes. The observation that measles virus (MeV) nucleoprotein (N) interacts with, and activates IRF-3, to induce CCL5 (also called RANTES), a pro-inflammatory cytokine, but not IFN-β [43], raises the issue of identifying the underlying molecular mechanism. In this chapter, there will be a short overview of MeV ability to induce the cellular innate responses, a more detailed review of IRF-3 activation and function, and a summary of the known MeV N and IRF-3 interactions. Finally, future areas of investigation will be proposed.

2. Measles Virus and the Cellular Innate Immune Response

MeV infection begins in the respiratory tract and disseminates to the whole body, mainly through the migration of infected monocytes and/or dendritic cells. The main physiopathological feature of this disease is the immunological paradox which associates a strong cellular immunosuppression and the elicitation of a life-long protective immunity against MeV re-infection (see [9] for review, see also Chapter 5). Virus infection elicits innate and adaptative immune response. Type I interferons, comprising the unique IFN-β gene and the multiple IFN-α , constitute an essential early component of the innate immune response. IFN-α/β confers to cells protection against viral infection via pleiotropic activities such as inhibition of protein synthesis and cell proliferation, enhancement of infected cell apoptosis and immunomodulation (reviewed in [2]). Type I interferon (IFN-α/β) and chemokines also provide the inflammatory link between innate and acquired immunity [2, 4]. MeV infection elicits innate immunity including both IFN response and secretion of numerous pro-inflammatory cytokines both *in vivo* and *in vitro*. After natural MeV infection, peripheral blood leucocytes from patients secrete IFN-α [5] although no circulating IFN-α in late samples has been reported [48]. Furthermore, plasma level of CXCL1 [17] and cerebrospinal fluid level of CXCL10 [37] are elevated in acute and chronic MeV infected patients, respectively. In humans vaccinated with an attenuated MeV strain, circulating IFN-α/β appears 3-4 days post infection (d.p.i.) to peak at 9-10 d.p.i., and fades thereafter [31]. *In vitro* MeV infection of human cells results in the secretion of type I IFN-α/β [25, 46, 47] and several chemokines such as CXCL1 [17], CCL2 [1, 26], CCL3 [20], CCL4 [40], CCL5 [1, 17], CCL7 [23] CCL8 [45], CCL19 [7], CCL20, CXCL8 (IL-8) [40], and CXCL10 [1, 26], but not CCL18 [41].

3. IRF-3 as a Key Player in the Cellular Innate Immune Response

IRF-3 is ubiquitously expressed as a stable latent transactivator of the cellular innate immune response. Virus infection primes the cellular innate immunity through the interaction of Pathogen Recognition Receptors (PRR) with Pathogen Associated Molecular Pattern (PAMP) (see [16] for review). These PRR are either cytosolic, as RNA helicases RIG-I and MDA-5, which are dedicated to sense PAMPS inside the cell, or transmembrane proteins expressed at the cell surface and/or in endosomes as Toll-like receptors (TLR), which detect extracellular PAMPS. Only the cytosolic pathway is relevant for description in the case of measles virus. Recognition of viral encoded dsRNA or 5' triphosphate ssRNA by MDA-5 [15] and RIG-I proteins [14, 32], respectively, results in the recruitment and activation of a mitochondrial anchored protein, herein referred as IPS-1 (see [12] for review). IPS-1 is the crossing starting point of three independent signaling pathways (Figure 1). (i) IPS-1 binds to and activates TBK1 kinase to phosphorylate IRF-3 or IRF-7. HSP90 [50] and cyclophilin B (CyB) [28] act as co-activators. Phosphorylated IRF-3 and IRF-7 homo- or heterodimerize and translocate into the nucleus where they can bind to interferon sensitive response elements (ISREs). (ii) IPS-1 recruits IKKγ and activate IKKα/IKKβ and the downstream NF-κB[p65-p50] heterodimer which translocates into the nucleus and can bind to NF-κB sites. (iii) The third pathway is less well defined and involves the ubiquitin ligase TRAF-6, likely a MAPKinase kinase and the JUNK kinase to activate the ATF-2/c-Jun (AP-1) heterodimer which, after translocation into the nucleus, binds to AP-1 sites. The promoter of the IFN-β gene consists in the close apposition of one AP1 site (also called positive regulatory domain IV or PRDIV), two ISRE sites (PRDIII and PRDI), and one NF-κB site (PRD-II). Maximal level of transcriptional synergy between AP-1 and NF-κB with IRF3/7 requires specific interactions with HMG I(Y) protein and the correct helical phasing of the binding sites of these proteins on the DNA helix [18, 51]. Together with the ability of IRF-3 to bind to both ATF2 and NF-κB p65, this coordinated binding to DNA results in the formation of a highly stable enhanceosome which binds to CREB binding protein (CBP) or P300 to ensure the recruitment of the RNA polymerase II complex.

4. The Set of Genes Activated by Phosphorylated IRF-3 Is Variable

In contrast to IFN-β promoter, the promoters of proinflammatory cytokine and IFN-α genes lack the AP-1 site and, sometimes, are even devoid of either ISRE or NF-κB site. In fact, the transactivating complex can be made of either homodimeric IRF-3 and heterodimeric

NF-κB[p65-p50] bound to the ISRE and NF-κB sites on DNA, or of a $[IRF\text{-}3]_2$/NF-κB[p65-p50] complex bound to a single NF-κB site or to a single ISRE site (Figure 1) (see [12] for review).

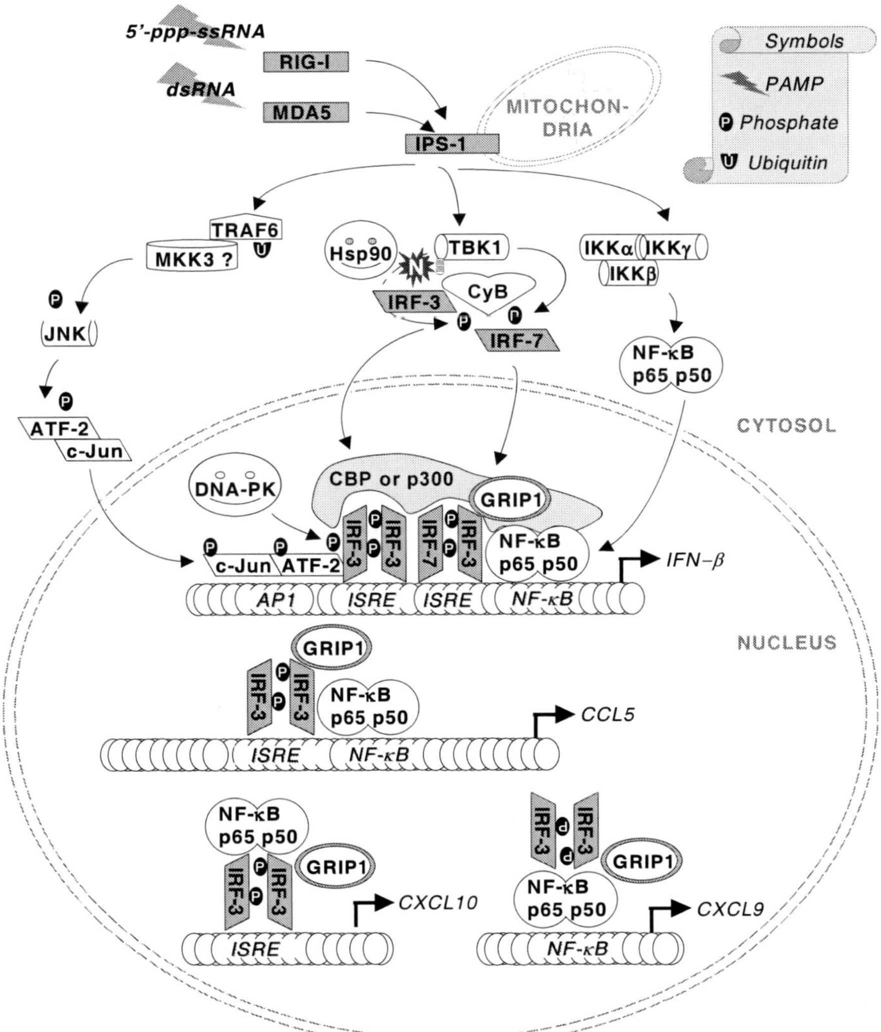

Figure 1. IRF-3 activation downstream to intracellular PRRs (adapted from [11, 12]). Upon recognition of viral derived 5'triphosphate ssRNA or dsRNA, RIG-I and MDA-5 interact with the mitochondrial IPS-1 and activates three sub-pathways leading to the formation of the IFN-β enhanceosome. The first leads to the activation of cJun/ATF-2 heterodimers which will bind to AP1 site: it is not well defined yet and likely involves TRAF-6, possibly MKK3, and JNK. The second leads to the phosphorylation by TBK1 and nuclear translocation of IRF-3 and IRF-7 homo- and hetero-dimers which bind to ISRE sites; HSP90 and CyB act as positive regulators of IRF-3 phosphorylation by TBK1. The third pathway allows the activation of NF-κB heterodimer through phosphorylation of I-κB by the IKKα/IKKβcomplex. Phosphorylation of IRF-3 by DNA-PK stabilises its binding to ISRE, and GRIP-1 tethers IRF-3 and NF-κB. The IRF-3/7 dimers recruit CBP/p300. The promoter regions of inflammatory cytokines are variable and often lack one (AP1) or two (ISRE or NF-κB) of the canonical enhanceosome binding sites; in these cases, IRF-3 homodimers can bind to ISRE and recruit NF-κB and reciprocally, NF-κB can bind to NF-κB binding sites and recruit IRF-3 homodimers. Putative step at which MeV N can act in order to activate IRF-3 (and NF-κB) is indicated by a black star.

In any case, IRF-3 also binds to the glucocorticoid receptor inhibitor protein GRIP1 which ensures a stable binding with NF-κB p65 and that of NF-κB[p65-p50] on DNA [35] (see [12] for review). ISRE is a general acronym characterised by a consensus DNA motif which comprises 4 groups with subtle variations in their sequences affecting their susceptibility to be transactivated by different complexes [24]:

(1) IRF-3 and/or IRF-7 dependent ISRE (PRDIII and PRDI) characteristic of the IFN-β promoter
(2) IRF-7 only dependent ISRE, called PRD-like element, present on the promoter of IFN-α genes
(3) IRF3 dependent or ISGF3 (interferon stimulated gene factor 3) dependent ISRE present on pro-inflammatory cytokine promoters. ISGF3 is a complex made of STAT1, STAT2 and IRF-9 and is induced by the activation of the IFN-α/β receptor (IFNAR)
(4) ISGF3 only dependent group: present on the promoters of IRF-7, and most of the antiviral effector genes called interferon stimulated gene (ISG) activated downstream to IFNAR activation by IFN-α/β binding [3]

Likewise, NF-κB binding sites share a common consensus sequence, but cofactor specificity for NF-κB dimers is determined by single nucleotide difference [21]. As a consequence, different combinations of ISRE and NF-κB sites, variable in number (1 or 2 sites) and in sequence, and additional sites, such as AP-1, explain the diversity pattern of gene expression during activation of cellular innate immunity according to the sets of transactivators that have been activated.

5. IRF-3 Activation by Measles Virus N

The seminal and unique observation that MeV N activates IRF-3 was made by J. Hiscott's group in 2001-2 [42, 43]. At 16 to 20 hours post-infection, activated phosphorylated IRF-3 at the key Ser^{385} and Ser^{386} residues was detected, and this form was able to bind to ISRE of ISG15 in complex with CBP *in vitro*. This activation requires active MeV transcription as shown by the lack of stimulation after exposure to UV-inactivated MeV or to a recombinant MeV lacking a functional L polymerase. Activation of IRF-3 was detected in the absence of any NF-κB activation. It was mimicked by the transient expression of MeV N protein alone or together with MeV-P protein. IRF-3 virus activated kinase (likely TBK-1 alone or in complex) could be co-immunoprecipitated with MeV N.

6. Measles Virus N_{tail} Binds to the IRF-3 IAD Region Where Binding Sites of Numerous Other Cell Partners Are Located

MeV N and IRF-3 can bind to each other through the interaction of a region within MeV N_{TAIL} (residues 415-523) with the IRF-3 interferon associated domain IAD (residues 198-394). Up to a dozen of cellular proteins have been identified so far to bind to IRF-3. The IRF-3 structure consists of a NH_2-terminal DNA binding domain (DBD) encompassing residues 1-111, a proline rich domain, the IAD and a COOH-terminal serine rich region (SRR), or regulatory domain (Figure 2).

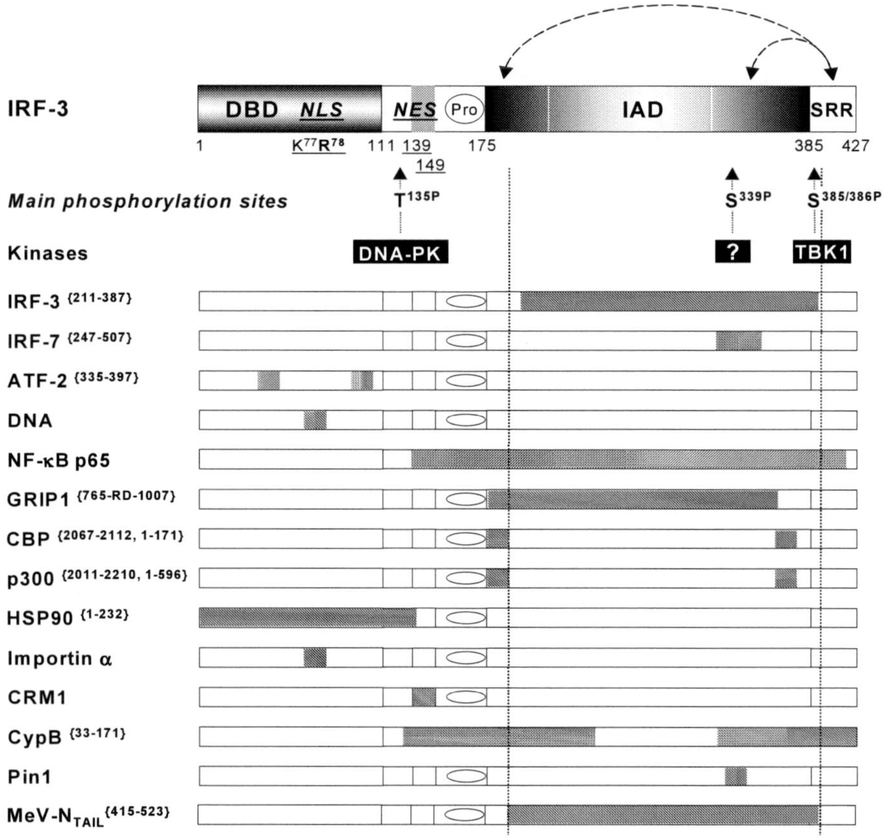

Figure 2. Human IRF-3 organization and molecular partners. DBD, DNA binding domain; NLS, nuclear localisation signal; NES, nuclear export signal; Pro, proline-rich region; IAD, interferon-associated domain; SRR, serine-rich region. Upper dotted arrows indicate auto-inhibitory interaction sites within IRF-3 [33]. Interacting sites on IRF-3 are: [37-51] and [98-101] for ATF-2 [30], [73-86] for DNA [30] and Importin α (Qip1, KPNA3) [19], [133-420] for NF-κB (p65RHD),[29], [188-369] for GRIP1 [35], [189-196] and [369-377] for CBP [34], [1-140] for HSP90 [50], [139-149] for CRM1 [19], [132-241] and [328-427] for cyclophilin B [28], [211-387] for homodimerisation, [328-357] for heterodimerisation with IRF-7 [22], [198-394] for N_{TAIL} [43], Phosphorylated S^{339} for Pin1 [36]. Two vertical dotted lines delimit the binding sites of N_{TAIL} which shows overlaps with binding sites of other molecular partners.

DBD contains a nuclear localization signal, NLS, which interacts with importin α subsets Qip1 and KPNA3 acting as an onward shuttle [19]. A nuclear export signal NES, which interacts with the chromosome maintenance 1 (CRM1) backward shuttle, is adjacent to the DBD. IRF-3 is the substrate of several serine-threonine kinases. TBK1 phosphorylates multiple sites such as $Ser^{402}Thr^{404}Ser^{405}$, $S^{396/398}$ and $Ser^{385/386}$ [44]. Phosphorylation of the latter is thought to relieve the auto-inhibitory interaction of SRR with two IAD regions as revealed by analysis of the crystal structure of IRF-3 [33] and allows IRF-3 homodimerisation of both IADs or heterodimerisation with IRF-7 IAD [22]. This activity is facilitated by HSP90 which binds to DBD of IRF-3 and to TBK1, to bring them dynamically into proximity and form an active substrate/enzyme complex [50]. CyB, a cis-trans peptidyl-prolyl isomerase, binds to IRF-3 IAD and is critically involved in its activation [28]. In the nucleus, IRF-3 holocomplex is stabilized by IRF-3 phosphorylation at position Thr^{135} by the DNA protein kinase DNA-PK. Phosphorylation of IRF-3 at the $Ser^{339}Pro^{340}$ motif, by a yet unknown kinase, targets IRF-3 for recognition by the peptidyl-prolyl isomerase Pin1 resulting in polyubiquitination and proteasome-dependent degradation of activated IRF-3 [36]. In the nucleus, activated homo- (and/or hetero-) dimerized IRF-3 can make different holocomplexes made of NF-κB[p65/p50], CBP/p300 and/or AP-1 c-Jun/ATF-2. Indeed, there are one ATF-2 binding site on IRF-3 DBD [30] and p65 or CBP/p300 binding sites located on IRF-3 IAD [29, 34]. IRF-3 has also its IAD simultaneously bound to the glucocorticoid receptor inhibitor 1 GRIP1. The binding of glucocorticoid onto its receptor GR results in the displacement of GRIP1 from IRF-3 to negatively regulate AP-1/NF-κB activated genes [35]. This mechanism provides a molecular support for the well-known anti-inflammatory activity of glucocorticoids.

7. Measles Virus N Activates (A Subset of) IRF-3 Dependent Proinflammatory Cytokine Genes but Not IFN-β

MeV N transcribed and translated from infectious MeV, or from a transfected eukaryotic vector, activates only a subset of IRF-3-driven genes. On one hand, IFN-β gene was not activated, likely because of the lack of NF-κB activation [43] and/or possibly to the lack of activation of IRF-7, which is necessary for optimal activation of IFN-β [13]. On the other hand, CCL5 and CCL3 genes were strongly activated [27, 43, 49]. It is unclear if the activation of CCL2, CCL4 [49] and CXCL8 genes [39] by UV-inactivated MeV is also related to MeV N mediated activation of IRF-3. Indeed for full activation of IRF-3, CCL5 and CCL3 genes, there is a strict requirement for neosynthesized N protein, since the passive delivery of UV-inactivated MeV nucleocapsid by virus particle fusion with target cells, or inhibition of protein synthesis, failed to result in a significant activation [27, 43, 49]. However, one cannot exclude that the MeV N assembled into nucleocapsids can activate a low and undetectable level of IRF-3, which would be sufficient for the activation of CCL2, CCL4 and/or CXCL8 genes, since nucleocapsids from vesicular stomatitis virus, another member of the *Mononegavirales* order, can efficiently activate IRF-3 [44]. The efficient activation of CCL5 gene by MeV N activated IRF-3 in the absence of NF-κB activation [43] is not expected, since efficient CCL5 gene activation requires activation of both of the IRF-3 and NF-κB pathways [8]. Therefore, one has to speculate that N-mediated IRF-3 activation results in a particular holocomplex strongly enough for CCL5 gene activation in the absence of NF-κB.

8. Underlying Molecular Mechanisms and Outcomes

In summary, MeV N protein, likely when in a monomeric form (N°) or within the N°P complex (see Chapter 1), interacts with IRF-3 to induce phosphorylation of the latter by TBK1 which leads to its homodimerisation and transactivation of a selective set of pro-inflammatory cytokines. This observation raises several issues both at the molecular and biological levels.

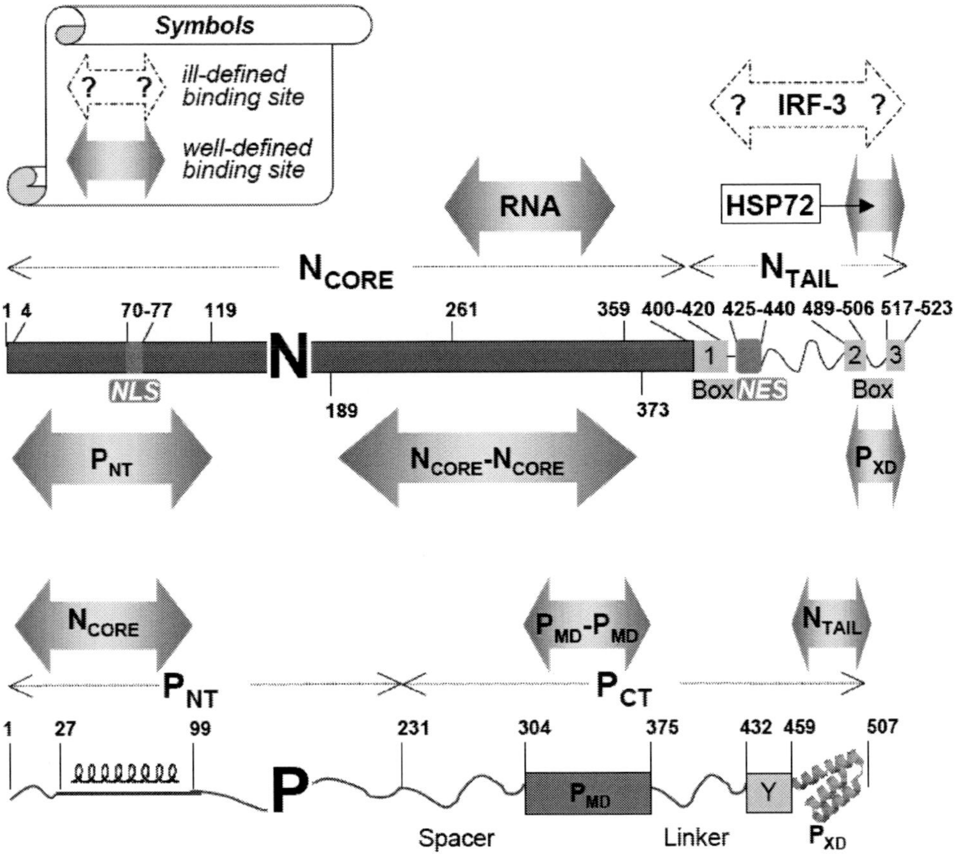

Figure 3. Schematic view of MeV nucleoprotein N and phosphoprotein P and localisation of the binding sites of their respective partners. The N protein exists in two forms, one monomeric and one polymerised. PNT and PCT are the natively unfolded and the structured domains of P protein, respectively. PCT is subdivided into P multimerisation domain (PMD), Y and XD domains separated by linkers (see Chapter 1 for details).

How the natively unfolded N_{TAIL} can convert the inactive IRF-3 into active transactivator? Does N_{TAIL}, alone or as whole N protein, act as a molecular scaffold, as does HSP90, to bridge TBK1 with its substrate IRF-3? Does MeV N also binds to TBK1 or to another TBK1 cellular partner? Indeed, the inability of MeV N to activate NF-κB [43] indicates that it acts downstream of IPS-1, which is the next common node for at least three different pathways of signal transduction. The apparent overlap of N_{TAIL} binding site on IRF-3 IAD with those of at least half a dozen of other cellular partners (see Figure 1 for details), as well as the apparent overlap of IRF-3 binding sites on MeV N_{TAIL} with those of its own viral and cellular partners (Figure 3), indicate (i) that a refined mapping of every binding site is required to identify possible competition amongst every set of partners for binding to the same target and (ii) that MeV N_{TAIL} interaction with IRF-3 should be timely regulated to ensure maintenance of efficient overall or partial IRF-3 and MeV N partnership and/or to endow them with novel functions. For example, both of the IRF-3 and MeV N have functional NLS and NES signals [38], and one can asked whether they are shuttling companions. Likewise, does the direct or indirect N recruitment of TBK1 enable this kinase to phosphorylate N protein or one of its partners such as the MeV P protein?

From the biological point of view, what can be the advantage for the virus to force expression of (a subset of) pro-inflammatory genes? Is it to enhance virus propagation in tissue? Intriguingly, during MeV infection, IRF-3 [42] and NF-κB [6, 10] are also activated and can lead to IFN-β activation [42, 46]. Therefore, what can be the biological advantage(s) of activating additional IRF-3 molecules?

In conclusion, the IRF-3 and MeV N partnership is a factual observation that the physiological role needs to be further deciphered. IRF-3 is a folded protein with at least 16 molecular partners and N_{TAIL} is a natively unfolded domain with at least half a dozen of molecular partners. How a single domain, either folded or unfolded, can accommodate, in a dynamically controlled fashion, the binding to so many molecular partners remains a puzzling area of investigation.

Acknowledgements

This work was supported in part by ANR-MIME. Florence Herschke was supported by a fellowship from MENRT.

References

[1] Berghall, H., J. Siren, D. Sarkar, I. Julkunen, P. B. Fisher, R. Vainionpaa, and S. Matikainen. 2006. The interferon-inducible RNA helicase, mda-5, is involved in measles virus-induced expression of antiviral cytokines. *Microbes Infect.* 8:2138-44.

[2] Bonjardim, C. A. 2005. Interferons (IFNs) are key cytokines in both innate and adaptive antiviral immune responses--and viruses counteract IFN action. *Microbes Infect.* 7:569-78.

[3] Brierley, M. M., and E. N. Fish. 2005. Stats: multifaceted regulators of transcription. J Interferon *Cytokine Res.* 25:733-44.

[4] Coelho, A. L., C. M. Hogaboam, and S. L. Kunkel. 2005. Chemokines provide the sustained inflammatory bridge between innate and acquired immunity. *Cytokine Growth Factor Rev.* 16:553-60.

[5] Crespi, M., J. K. Struthers, A. N. Smith, and S. F. Lyons. 1988. Interferon status after measles virus infection. *S. Afr. Med. J.* 73:711-2.

[6] Dhib-Jalbut, S., J. Xia, H. Rangaviggula, Y. Y. Fang, and T. Lee. 1999. Failure of measles virus to activate nuclear factor-kappa B in neuronal cells: implications on the immune response to viral infections in the central nervous system. *J. Immunol.* 162:4024-9.

[7] Dubois, B., P. J. Lamy, K. Chemin, A. Lachaux, and D. Kaiserlian. 2001. Measles virus exploits dendritic cells to suppress CD4+ T-cell proliferation via expression of surface viral glycoproteins independently of T-cell trans-infection. *Cell Immunol.* 214:173-83.

[8] Genin, P., M. Algarte, P. Roof, R. Lin, and J. Hiscott. 2000. Regulation of RANTES chemokine gene expression requires cooperativity between NF-kappa B and IFN-regulatory factor transcription factors. *J. Immunol.* 164:5352-61.

[9] Gerlier, D., H. Valentin, D. Laine, C. Rabourdin-Combe, and C. Servet-Delprat. 2006. Subversion of the immune system by measles virus: a model for the intricate interplay between a virus and the human immune system, p. 225-292. *In* P. J. Lachman and M. B. A. Oldstone (ed.), *Microbial Subversion of Host Immunity.* Caister Academic Press, Norwalk, UK.

[10] Helin, E., R. Vainionpaa, T. Hyypia, I. Julkunen, and S. Matikainen. 2001. Measles virus activates NF-kappa B and STAT transcription factors and production of IFN-alpha/beta and IL-6 in the human lung epithelial cell line A549. *Virology* 290:1-10.

[11] Honda, K., A. Takaoka, and T. Taniguchi. 2006. Type I inteferon gene induction by the interferon regulatory factor family of transcription factors. *Immunity* 25:349-60.

[12] Honda, K., and T. Taniguchi. 2006. IRFs: master regulators of signalling by Toll-like receptors and cytosolic pattern-recognition receptors. *Nat. Rev. Immunol.* 6:644-58.

[13] Honda, K., H. Yanai, H. Negishi, M. Asagiri, M. Sato, T. Mizutani, N. Shimada, Y. Ohba, A. Takaoka, N. Yoshida, and T. Taniguchi. 2005. IRF-7 is the master regulator of type-I interferon-dependent immune responses. *Nature* 434:772-7.

[14] Hornung, V., J. Ellegast, S. Kim, K. Brzozka, A. Jung, H. Kato, H. Poeck, S. Akira, K. K. Conzelmann, M. Schlee, S. Endres, and G. Hartmann. 2006. 5'-Triphosphate RNA Is the Ligand for RIG-I. *Science*.

[15] Kato, H., O. Takeuchi, S. Sato, M. Yoneyama, M. Yamamoto, K. Matsui, S. Uematsu, A. Jung, T. Kawai, K. J. Ishii, O. Yamaguchi, K. Otsu, T. Tsujimura, C. S. Koh, C. Reis e Sousa, Y. Matsuura, T. Fujita, and S. Akira. 2006. Differential roles of MDA5 and RIG-I helicases in the recognition of RNA viruses. *Nature* 441:101-5.

[16] Kawai, T., and S. Akira. 2006. Innate immune recognition of viral infection. *Nat. Immunol.* 7:131-7.

[17] Kim, H. S., W. D. Kim, and Y. H. Lee. 2003. Production and expression of Gro-alpha and RANTES by peripheral blood mononuclear cells isolated from patients with Kawasaki disease and measles. *J. Korean Med. Sci.* 18:381-6.

[18] Kim, T. K., and T. Maniatis. 1997. The mechanism of transcriptional synergy of an in vitro assembled interferon-beta enhanceosome. *Mol. Cell* 1:119-29.

[19] Kumar, K. P., K. M. McBride, B. K. Weaver, C. Dingwall, and N. C. Reich. 2000. Regulated nuclear-cytoplasmic localization of interferon regulatory factor 3, a subunit of double-stranded RNA-activated factor 1. *Mol. Cell Biol.* 20:4159-68.

[20] Kutza, J., L. Crim, S. Feldman, M. P. Hayes, M. Gruber, J. Beeler, and K. A. Clouse. 2000. Macrophage colony-stimulating factor antagonists inhibit replication of HIV-1 in human macrophages. *J. Immunol.* 164:4955-60.

[21] Leung, T. H., A. Hoffmann, and D. Baltimore. 2004. One nucleotide in a kappaB site can determine cofactor specificity for NF-kappaB dimers. *Cell* 118:453-64.

[22] Lin, R., Y. Mamane, and J. Hiscott. 2000. Multiple regulatory domains control IRF-7 activity in response to virus infection. *J. Biol. Chem.* 275:34320-7.

[23] Menten, P., P. Proost, S. Struyf, E. Van Coillie, W. Put, J. P. Lenaerts, R. Conings, J. M. Jaspar, D. De Groote, A. Billiau, G. Opdenakker, and J. Van Damme. 1999. Differential induction of monocyte chemotactic protein-3 in mononuclear leukocytes and fibroblasts by interferon-alpha/beta and interferon-gamma reveals MCP-3 heterogeneity. *Eur. J. Immunol.* 29:678-85.

[24] Nakaya, T., M. Sato, N. Hata, M. Asagiri, H. Suemori, S. Noguchi, N. Tanaka, and T. Taniguchi. 2001. Gene induction pathways mediated by distinct IRFs during viral infection. *Biochem. Biophys. Res. Commun.* 283:1150-6.

[25] Naniche, D., A. Yeh, D. Eto, M. Manchester, R. M. Friedman, and M. B. Oldstone. 2000. Evasion of host defenses by measles virus: wild-type measles virus infection interferes with induction of Alpha/Beta interferon production. *J. Virol.* 74:7478-84.

[26] Nazar, A. S., G. Cheng, H. S. Shin, P. N. Brothers, S. Dhib-Jalbut, M. L. Shin, and P. Vanguri. 1997. Induction of IP-10 chemokine promoter by measles virus: comparison with interferon-gamma shows the use of the same response element but with differential DNA-protein binding profiles. *J. Neuroimmunol.* 77:116-27.

[27] Noe, K. H., C. Cenciarelli, S. A. Moyer, P. A. Rota, and M. L. Shin. 1999. Requirements for measles virus induction of RANTES chemokine in human astrocytoma-derived U373 cells. *J. Virol.* 73:3117-24.

[28] Obata, Y., K. Yamamoto, M. Miyazaki, K. Shimotohno, S. Kohno, and T. Matsuyama. 2005. Role of cyclophilin B in activation of interferon regulatory factor-3. *J. Biol. Chem.* 280:18355-60.

[29] Ogawa, S., J. Lozach, C. Benner, G. Pascual, R. K. Tangirala, S. Westin, A. Hoffmann, S. Subramaniam, M. David, M. G. Rosenfeld, and C. K. Glass. 2005. Molecular determinants of crosstalk between nuclear receptors and toll-like receptors. *Cell* 122:707-21.

[30] Panne, D., T. Maniatis, and S. C. Harrison. 2004. Crystal structure of ATF-2/c-Jun and IRF-3 bound to the interferon-beta enhancer. *Embo J.* 23:4384-93.

[31] Petralli, J. K., T. C. Merigan, and J. R. Wilbur. 1965. Circulating Interferon After Measles Vaccination. *N. Engl. J. Med.* 273:198-201.

[32] Pichlmair, A., O. Schulz, C. P. Tan, T. I. Naslund, P. Liljestrom, F. Weber, and E. S. C. Reis. 2006. RIG-I-Mediated Antiviral Responses to Single-Stranded RNA Bearing 5' Phosphates. *Science.*

[33] Qin, B. Y., C. Liu, S. S. Lam, H. Srinath, R. Delston, J. J. Correia, R. Derynck, and K. Lin. 2003. Crystal structure of IRF-3 reveals mechanism of autoinhibition and virus-induced phosphoactivation. *Nat. Struct. Biol.* 10:913-21.

[34] Qin, B. Y., C. Liu, H. Srinath, S. S. Lam, J. J. Correia, R. Derynck, and K. Lin. 2005. Crystal structure of IRF-3 in complex with CBP. *Structure* 13:1269-77.

[35] Reily, M. M., C. Pantoja, X. Hu, Y. Chinenov, and I. Rogatsky. 2006. The GRIP1:IRF3 interaction as a target for glucocorticoid receptor-mediated immunosuppression. *Embo J.* 25:108-17.

[36] Saitoh, T., A. Tun-Kyi, A. Ryo, M. Yamamoto, G. Finn, T. Fujita, S. Akira, N. Yamamoto, K. P. Lu, and S. Yamaoka. 2006. Negative regulation of interferon-regulatory factor 3-dependent innate antiviral response by the prolyl isomerase Pin1. *Nat. Immunol.* 7:598-605.

[37] Saruhan-Direskeneli, G., C. Gurses, V. Demirbilek, S. P. Yentur, G. Yilmaz, E. Onal, Z. Yapici, C. Yalcinkaya, O. Cokar, G. Akman-Demir, and A. Gokyigit. 2005. Elevated interleukin-12 and CXCL10 in subacute sclerosing panencephalitis. *Cytokine* 32:104-10.

[38] Sato, H., M. Masuda, R. Miura, M. Yoneda, and C. Kai. 2006. Morbillivirus nucleoprotein possesses a novel nuclear localization signal and a CRM1-independent nuclear export signal. *Virology* 352:121-30.

[39] Sato, H., R. Miura, and C. Kai. 2005. Measles virus infection induces interleukin-8 release in human pulmonary epithelial cells. *Comp. Immunol. Microbiol. Infect. Dis.* 28:311-20.

[40] Schutyser, E., S. Struyf, P. Menten, J. P. Lenaerts, R. Conings, W. Put, A. Wuyts, P. Proost, and J. Van Damme. 2000. Regulated production and molecular diversity of human liver and activation-regulated chemokine/macrophage inflammatory protein-3 alpha from normal and transformed cells. *J. Immunol.* 165:4470-7.

[41] Schutyser, E., S. Struyf, A. Wuyts, W. Put, K. Geboes, B. Grillet, G. Opdenakker, and J. Van Damme. 2001. Selective induction of CCL18/PARC by staphylococcal enterotoxins in mononuclear cells and enhanced levels in septic and rheumatoid arthritis. *Eur. J. Immunol.* 31:3755-62.

[42] Servant, M. J., B. ten Oever, C. LePage, L. Conti, S. Gessani, I. Julkunen, R. Lin, and J. Hiscott. 2001. Identification of distinct signaling pathways leading to the phosphorylation of interferon regulatory factor 3. *J. Biol. Chem.* 276:355-63.

[43] tenOever, B. R., M. J. Servant, N. Grandvaux, R. Lin, and J. Hiscott. 2002. Recognition of the measles virus nucleocapsid as a mechanism of IRF-3 activation. *J. Virol.* 76:3659-3669.

[44] tenOever, B. R., S. Sharma, W. Zou, Q. Sun, N. Grandvaux, I. Julkunen, H. Hemmi, M. Yamamoto, S. Akira, W. C. Yeh, R. Lin, and J. Hiscott. 2004. Activation of TBK1 and IKKvarepsilon kinases by vesicular stomatitis virus infection and the role of viral ribonucleoprotein in the development of interferon antiviral immunity. *J. Virol.* 78:10636-49.

[45] Van Damme, J., P. Proost, W. Put, S. Arens, J. P. Lenaerts, R. Conings, G. Opdenakker, H. Heremans, and A. Billiau. 1994. Induction of monocyte chemotactic proteins MCP-1 and MCP-2 in human fibroblasts and leukocytes by cytokines and

cytokine inducers. Chemical synthesis of MCP-2 and development of a specific RIA. *J. Immunol.* 152:5495-502.

[46] Vidalain, P. O., D. Laine, Y. Zaffran, O. Azocar, C. Servet-Delprat, T. F. Wild, C. Rabourdin-Combe, and H. Valentin. 2002. Interferons mediate terminal differentiation of human cortical thymic epithelial cells. *J. Virol.* 76:6415-24.

[47] Volckaert-Vervliet, G., H. Heremans, M. De Ley, and A. Billiau. 1978. Interferon induction and action in human lymphoblastoid cells infected with measles virus. *J. Gen. Virol.* 41:459-66.

[48] Wheelock, E. F., and W. A. Sibley. 1964. Interferon In Human Serum During Clinical Viral Infections. *Lancet* 128:382-5.

[49] Xiao, B. G., A. Mousa, P. Kivisakk, A. Seiger, M. Bakhiet, and H. Link. 1998. Induction of beta-family chemokines mRNA in human embryonic astrocytes by inflammatory cytokines and measles virus protein. *J. Neurocytol.* 27:575-80.

[50] Yang, K., H. Shi, R. Qi, S. Sun, Y. Tang, B. Zhang, and C. Wang. 2006. Hsp90 regulates activation of interferon regulatory factor 3 and TBK-1 stabilization in Sendai virus-infected cells. *Mol. Biol. Cell* 17:1461-71.

[51] Yie, J., M. Merika, N. Munshi, G. Chen, and D. Thanos. 1999. The role of HMG I(Y) in the assembly and function of the IFN-beta enhanceosome. *Embo J.* 18:3074-89.

In: Measles Virus Nucleoprotein
Editors: S. Longhi, pp. 113-152

ISBN: 978-1-60021-629-9

Chapter V

Interaction of Measles Virus Nucleoprotein with Cell Surface Receptors: Impact on Cell Biology and Immune Response

David Laine[1,2], Pierre-Olivier Vidalain[1,3], Arije Gahnnam[4], Jean-Claude Cortay[4], Denis Gerlier[4], Chantal Rabourdin-Combe[1] and Hélène Valentin[1*]

[1]Laboratoire d'Immunobiologie Fondamentale et Clinique - INSERM U503 and UCBL1 IFR128 BioSciences Lyon-Gerland
21 Avenue Tony Garnier - 69365 Lyon Cedex 07 - France
[2] The Walter and Eliza Hall Institute of Medical Research - Autoimmunity and Transplantation Division - 1G Royal Parade
Parkville - 3050 - Victoria - Australia
[3] Present address: Laboratoire de Génomique Virale et Vaccination, CNRS URA1930, Institut Pasteur, 28 rue du Dr. Roux, 75015 Paris, France
[4] Virologie et Pathogenèse Virale - CNRS - Université de Lyon 1
UMR5537 - IFR 62 Laennec - 69372 Lyon Cedex 08 - France

Abstract

The major physiopathological feature of measles virus (MeV) infection is the induction of a specific long lasting anti-viral immune response, which coincides with the appearance of a profound and transient immunosuppression. On one hand, the adaptive immune response efficiency is illustrated by the clearance of the infection within 2 weeks and by the long-life protection against reinfection. On the other hand, the immunosuppression is characterized by a dramatic impairment of humoral and cellular

* Corresponding author : Laboratoire d'Immunobiologie Fondamentale et Clinique - INSERM U503 and UCBL1 IFR128 BioSciences Lyon-Gerland. 21 Avenue Tony Garnier. 69365 Lyon Cedex 07 – France. Phone: (33) 4 37 28 23 51. Fax: (33) 4 37 28 23 41. E-mail: valentin@cervi-lyon.inserm.fr

immune responses to unrelated antigens. This feature is induced by direct infection and indirect mechanisms, such as alteration of the biology of uninfected cells by viral proteins. The surprising observation, that MeV nucleoprotein (N) antibodies are the first to be produced in high amounts after MeV infection provides the basis for the hypothesis that the cytosolic MeV N could be released in the extracellular milieu and therefore interact with various extracellular partners with a significant impact on immune responses. This unforeseen role of MeV N involves a dual activity in the development of both MeV-specific immune response and immunosuppression. The analysis of MeV N in both *in vitro* and *in vivo* experimental models, as well as in human patients with measles revealed the complexity of the mechanism of action of this protein.

List of Abbreviations

ADCC	Antibody dependent cell cytotoxicity
Antibody	Ab
Ag	Antigen
APCs	Antigen presenting cells
BCR	B cell receptor
CDV	Canine distemper virus
CTL	Cytotoxic T lymphocyte
DCs	Dendritic cells
DTH	Delayed-type hypersensitivity
F	Fusion protein
FcR	Fc receptor
FcγRII	Immunoglobulin G Fc receptor II
FDCs	Follicular dendritic cells
H	Hemagglutinin
HIV-I	Human immunodeficiency virus type I
hsp72	Major inducible 70 kDa heat shock protein
HSR	Hypersensitivity response
Ig	Immunoglobulin
IL	Interleukin
ITAM	Immunotyrosine based activatory motif
ITIM	Immunotyrosine based inhibitory motif
L	Large protein
mAb	Monoclonal antibody
MGCs	Multinucleated giant cells
MHC	Major histocompatibility complex
MeV	Measles virus
MFI	Mean fluorescence intensity
N	Nucleoprotein
NC	Nucleocapsid
N_{CORE}	N-terminal moiety of N (aa 1-400)
NK	Natural killer

NR	Nucleoprotein receptor
N_{TAIL}	C-terminal moiety of N (aa 401-525)
P	Phosphoprotein
PAMPs	Pathogen-associated molecular pattern
PBLs	Peripheral blood lymphocytes
PBMCs	Peripheral blood mononuclear cells
PCT	Phosphoprotein C-terminal domain
PNT	Phosphoprotein N-terminal domain
PPRV	Peste des petits ruminants virus
RNA	Ribonucleic acid
RPV	Rinderpest virus
TCR	T cell receptor
TEC	Thymic epithelial cell
T_H	T helper
TLR	Toll like receptor
TRAIL	TNF-related apoptosis-inducing ligand
TRAIL-R	TRAIL receptor
TREC	TCR rearrangement excision circles
WFCs	Warthin-Finkeldey cells
WHO	World Health Organization

1. Introduction

Measles virus (MeV) is responsible for an acute childhood disease with common outbreaks in underdeveloped countries where there is lower socioeconomic status and low access to healthcare. Natural infection occurs through the respiratory tract and is restricted to human, although transmission to some monkey species can occur. All primary MeV infections give rise to the disease. The initial symptoms are abrupt cough, fever and inflamed conjunctivae and then the appearance of Koplik spots in the mouth. The rash appears approximately three days after the initial symptoms on the head, and progressively spreads to eventually cover the entire body. The viremia occurs within the 10-14 days latent period to culminate with the 2-3 day rash and outcome of the clinical signs (Figure 1). The virus is cleared after about 20 days thanks to the induction of an efficient humoral and cellular immunity against MeV (Figure 1). Paradoxically, MeV is also characterized by a profound cellular immunosuppression favouring opportunistic secondary infections that could be observed after the rash and might be life-threatening (Figure 1). The immunosuppression is characterized by ablation of tuberculinic test-delayed-type hypersensitivity (DTH) [117], severe lymphopenia, impairment of lymphocyte proliferation from MeV infected patients when *in vitro* stimulated using polyclonal activators [52, 57], impairment of antibody (Ab) response to typhoid vaccination [122], and a shift towards a type 2 helper T cell (T_H2) response [45]. Many studies have revealed the pivotal role of MeV proteins, especially MeV nucleoprotein (N) in the initiation of the MeV immune response as well as in the development of MeV-induced immunosuppression.

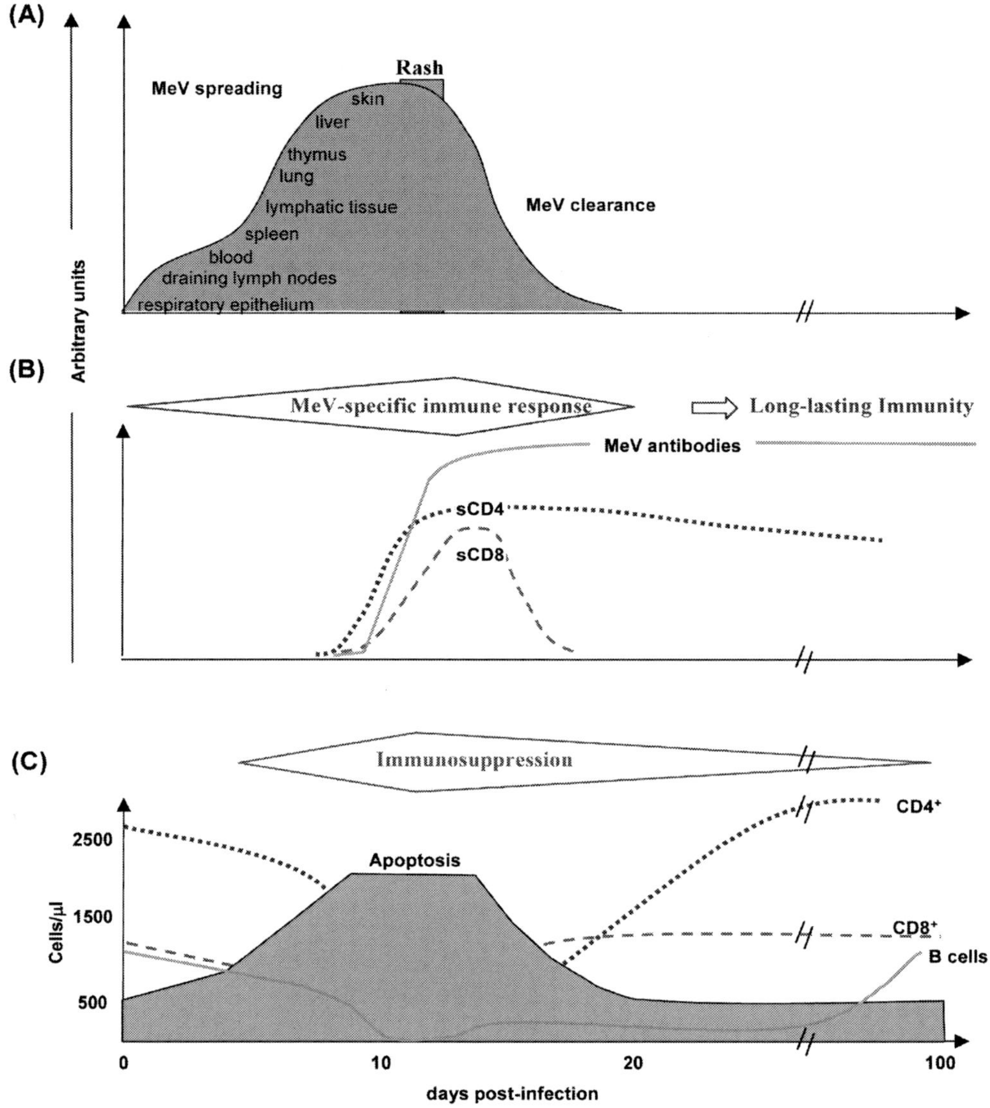

Figure 1. Overview of the time course of the immune response in MeV-infected patients. (A) MeV infects humans via the respiratory track before spreading to the draining lymph nodes, blood and tissues. When MeV reaches the skin at 7 days post-infection, the characteristic rash appears. MeV clearance occurs soon after the rash appearance. (B) MeV specific immune response is induced within the first week of MeV infection and gives a life-long protection against reinfection. Both cellular and humoral responses are induced, as assessed by the presence of soluble CD4/CD8 in plasma of infected patients and the production of MeV-specific Abs. (C) MeV-induced immunosuppression is detectable at the onset of the rash and can last for 100 days post-infection. The hallmark of this depression is a severe T and B lymphopenia. Note that CD8+ T cells are the first population to recover, followed by CD4+ T cells and eventually by B cells.

In this review, we will focus on *(i)* the mechanisms involved in extracellular MeV N release, *(ii)* the characterization of MeV N extracellular receptors, *(iii)* MeV N-mediated efficient immune response and immunosuppression. We will finally propose a model to explain how MeV N-mediated induction of immune response cohabits with MeV N-induced immunosuppression. Each section will begin with a brief introduction on innate/adaptive immunity responses against MeV before focusing on the specific roles of MeV N on immune response modulation.

2. Different Pathways Leading to MeV N Release

MeV N packs the viral RNA genome to form a helical nucleocapsid (NC) onto which the phosphoprotein (P) and the large (L) protein are bound to form a ribonucleoproteic complex (Figure 2A and B). This complex protects the viral RNA in the virion and also constitutes the template for the transcription and replication of the viral genome within the cytosol of infected cells (see Chapter 1). Thus, MeV N was thought to be an internal protein, located within either viral particles or infected cells. However, recent findings indicate that significant amounts of MeV N and/or NC can be found in the extracellular compartment and thus can interact with specific extracellular receptors (Figure 2C). The evidence of extracellular MeV N release came from the in vivo characterization of a strong humoral anti-N immune response, which is first detectable at the onset of the rash [10, 79]. Indeed, B cells are activated after B cell receptor (BCR) cross-linking by a specific extracellular antigen (Ag), unlike T cells, that require Ag to be degraded and presented at the surface of an Ag-presenting cell (APC) in the context of major histocompatibility complex (MHC) (see section 3).

The major source of extracellular MeV N is more likely from cytosolic origin, since MeV N is both the first and the more abundant MeV viral protein to be synthesized in infected cells [93]. The mechanisms responsible for the release of MeV N out of infected cells are complex (Figure 3). They involve three different, yet complementary, pathways including cell surface delivery and secretion into the extracellular milieu after cell death.

The first mechanism involves the release of MeV N from infected cells after cell apoptosis, necrosis and/or cytolysis [67]. MeV strains with different cytopathic activities revealed a strict correlation between MeV replication, MeV-induced apoptosis and the amount of MeV N release in the extracellular medium (Figure 3A). The exact mechanism leading to MeV N release by apoptotic cells has not been identified yet. One possibility is that MeV-induced cell death leads to the rupture of cell membranes, thus making cytosolic MeV N available for extracellular partners. Interestingly, degenerative and/or necrotic lesions of lymphoid organ are a hallmark of MeV infection [106]. They include the formation of syncytia or multinucleated giant cells (MGCs) such as Warthin-Finkeldey cells (WFCs) within the germinal centers and the follicular areas of the draining lymph nodes [80]. Syncytium formation is proportional to MeV replication and syncytia induce and amplify apoptosis [98]. Thus, mechanisms boosting MeV replication, e.g. CD40 activation, increase syncytium formation, and intracellular N synthesis (for a review, see [101]). By interacting with N_{TAIL}, hsp72 also stimulates the transcription activity of MeV and syncytium formation ([82, 112, 128], see also Chapter 3). Altogether, these combined mechanisms result in the accumulation of high amounts of intracellular N, which are released in the extracellular compartment after apoptosis, necrosis and/or cytolysis.

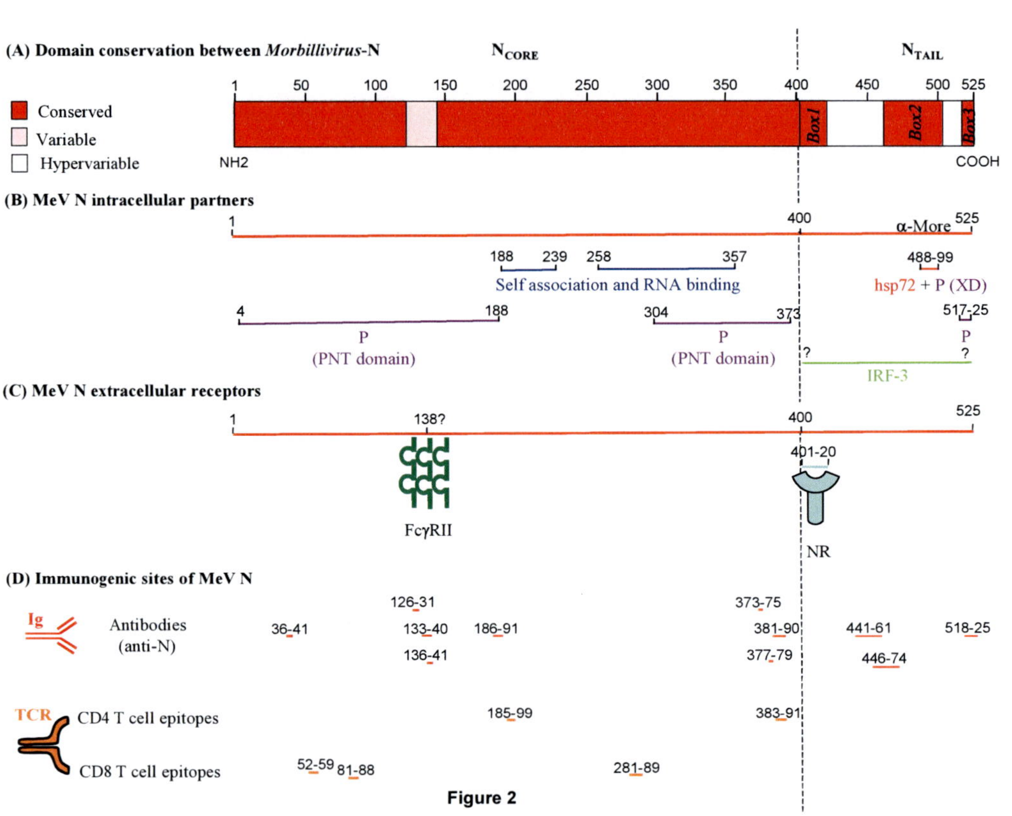

Figure 2. Schematic diagram of MeV N immunodominant epitopes and regions involved in binding to its cellular and extracellular partners. (A) Two regions can be distinguished according to the sequence comparison between Morbillivirus N: a conserved N_{CORE} domain (residues 1-400) and a hypervariable N_{TAIL} domain (residues 401-525). N_{CORE} has one variable region (residues 122-144) and N_{TAIL} has three conserved regions (residues 401–420, 489–506 and 517–525, corresponding to Box1, 2 and 3 respectively). The location of the N_{CORE}-PNT, N_{CORE}-N_{CORE} and N_{CORE}-RNA binding sites, as reported by [7],[60] is shown. (B) N_{CORE} can spontaneously self-assemble on RNA to form NCs (or NC-like particles), whereas N_{TAIL} contains binding site to P and hsp72. The helical molecular recognition element (α-MoRE, residues 488–99) is involved in binding to both P (XD) and hsp72 [16]. (C) MeV N interacts with three extracellular cell-surface receptors: BCR, FcγRII (CD32) and a yet uncharacterized receptor referred to as N receptor (NR). We postulate that residue 138 is crucial for binding to FcγRII. NR binds N_{TAIL} via Box 1. (D) Immunodominant MeV N epitopes are characterized by induction of either mAbs or MeV N-specific T cell clones. N-specific Abs mainly recognize sequences from N_{CORE} and, to a lower extent, N_{TAIL}. To our knowledge, $CD4^+$ and $CD8^+$ T-cell responses only target the N_{CORE} domain, which is less sensitive to proteolytic degradation.

This led us to hypothesize that MeV N is released within the lymphoid organs, and, after binding to cell surface receptors, acts locally in B and T cell areas.

Secondly, significant amount of MeV N, but not of other cytosolic MeV proteins, can be detected at the surface of viable MeV-infected cells [67, 73], suggesting a specific MeV N delivery to the cell membranes. The proposed model for delivery of MeV to the plasma membrane implies translocation of the native MeV N into the endo-lysosomal compartment, where it interacts with Fc receptors (FcR; see section 3) prior to its exportation to the cell surface and/or extracellular release [63, 73] (Figure 3B).

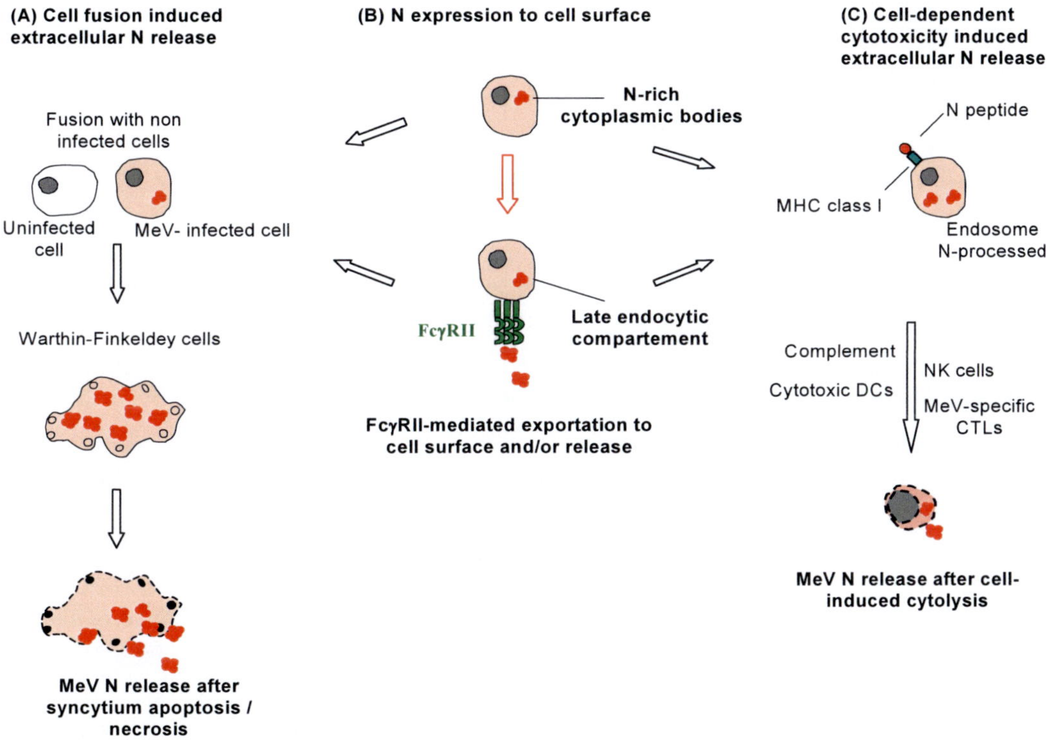

Figure 3. Multiple pathways for MeV N extracellular release . (A) MeV infected cells can fuse with uninfected cells to form syncytia (also called Warthin-finkeldey cells) in vivo. These giant multinucleated cells will eventually die and release N in the extracellular milieu by apoptosis and/or secondary necrosis. (B) Infected cells express cytosolic N in high quantities. Cells can directly export cytosolic N to cell surface via the endocytic compartment via FcγRII. (C) Infected cells can express at the cell surface the viral F glycoprotein and/or viral peptides bound to their MHC-class I molecules. This will activate the complement or cytotoxic cells (NK, DCs or CD8+ T cells). As a result, infected cells are lysed and cytosolic N is released into the extracellular compartment.

Finally, apoptosis and/or necrosis of MeV-infected cells can also be mediated in vivo by the immune system, including the complement, natural killer (NK) cells, cytotoxic dendritic cells (DCs) and $CD8^+$ T cells (Figure 3C). MeV F expressed on MeV-infected cells activates the alternative pathway of the complement [26], which can lead to cell lysis [103]. NK cells constitute a major component of the innate immune system, and they were named natural killer because they do not require any Ag-specific priming in order to kill cells that have low levels of MHC-class I cell surface marker molecules, a situation which could arise upon viral infection. NK cells are cytotoxic through the release of small granules containing perforin and granzymes. Although there is no direct evidence of NK activation early after MeV infection, NK cells are increased in number and activated just after the onset of the rash [84]. The second immune cell population possibly involved in MeV N release from infected cells is the cytotoxic DCs. In vitro differentiated DCs, once infected by MeV, exhibited TNF-related apoptosis-inducing ligand (TRAIL)-dependent cytotoxicity [113, 114]. TRAIL is a member of the TNF family, closely related to three death-inducing ligands: FasL, TNF-α, and

TWEAK/Apo3L. TRAIL may have a pivotal function in antiviral immune responses since cells infected by a virus are sensitized to TRAIL-induced apoptosis by yet unidentified mechanisms. Finally, $CD8^+$ T cells are also referred to as cytotoxic T lymphocytes (CTL) because they can recognize viral peptides on the surface of infected cells in the context of MHC-class I molecules to specifically kill infected cells. CTL can induce apoptosis by cell-cell interaction (i.e. Fas-Fas ligand interaction) or by secreting cytotoxins such as perforin and granzymes. Markers of CD8 activation are detected at the onset of the rash, strongly suggesting that CTL are an important component of the anti-MeV immune response (see section 4). Thus, cytotoxic immune cells are presumed to be additionally involved in MeV N release in the extracellular compartment by inducing MeV-infected cell apoptosis.

As previously described, intracellular MeV N can be distinguished into either a cytosolic or nuclear pool (see Chapter 1). Nuclear MeV N is the nascent protein and is not associated with P and constitutes a minor pool compared to cytolosic MeV N, which is associated with P. Whether both cytosolic and nuclear MeV N are released is yet to be determined. The release of MeV N into the extracellular milieu implies that MeV N becomes accessible to its cellular receptor(s), expressed by non-infected cells. Alternatively, cell surface-bound MeV N on infected cells may also directly interact with neighbouring cells to modulate their biology. As MeV N is highly sensible to serum protease, it favours a local role of MeV N within the organs where it has been released. Next, we will focus on the consequences of MeV N release upon the modulation of immune response after binding to its extracellular receptors.

3. Three Cell Surface Receptors for Native MeV N

To date, native MeV N is thought to bind to three different cell surface receptors: specific BCR expressed on N-specific B lymphocytes [43], low-affinity FcR for IgG of type II (FcγRII, CD32) [91], and a yet unidentified protein receptor called nucleoprotein receptor (NR) [67].

3.1. N-Specific B Cell Receptor (BCR)

The BCR is a multiprotein complex containing the heavy chain membrane-bound immunoglobulin (Ig) molecules and disulfide-linked heterodimers of Igα and Igβ. Within each of the Igα and Igβ cytoplasmic tails is embedded a consensus sequence, the immunoreceptor tyrosine-based activation motif (ITAM) D/E-x(7)-D/E-x(2)-Y-x(2)-L-x(7)-Y-x(2)-I/L (x denotes any amino acid), which is common to other multichain immune recognition receptor subunits, including three CD3 chains associated to T cell receptor (TCR) and certain Fc receptors (FcR), such as FcγRIII [94]. Upon BCR aggregation by its native ligand (i.e. MeV N), the two tyrosines of the ITAM motif become phosphorylated and drive signalling cascades for B cell activation and eventually for further differentiation into specific effectors (plasma cells) or memory cells. The fact that the most abundant and rapidly produced antibodies (Abs) are directed against MeV N implies that native extracellular MeV N binds specifically to BCR. Although the MeV N Ab isotype classes have been extensively studied *in vitro* (see section 4), little is known about the role of MeV N Abs in the immune response. To date, distinct Agic determinants of the MeV N were localized using monoclonal Abs (mAbs): 8 B-specific epitopes have been mapped to the N-terminal region, N_{CORE} (residues 36-41, 126-131, 133-140, 181-191, 373-375, 377-379 and 381-390) and only 3 epitopes in the C-terminal region of the protein N_{TAIL} (residues 441-461, 466-474 and 518-525) [18, 33, 39] (Figure 2D).

3.2. Low-Affinity Fc Receptor for IgG of Type II (FcγRII, Cd32)

The transmembrane FcγRIIs are low affinity receptors for the Fc portion of IgG Abs that bind immune complexes with a high avidity but are unable to bind circulating IgG, even in high concentrations. FcγRIIs are found on both myeloid and lymphoid cells, contrary to the FcγRI and FcγRIII which are expressed primarily on cells of the myeloid lineage and can mediate effector functions such as phagocytosis. Human FcγRIIs (FcγRIIa1-2, FcγRIIb1-3, and FcγRIIc1-4) represent alternatively spliced forms of tree genes, differing by their cytoplasmic tails (see [92, 108] for review) (Figure 4). The FcγRIIc gene represents an unequal cross-over event between genes IIa and IIb [120]. Contrary to humans, only two isoforms FcγRIIB1 and FcγRIIB2 are expressed in mouse.

Nearly all circulating conventional DCs and a major population of monocyte-derived DCs express both FcγRIIA and FcγRIIB2 [14]. Similarly, monocytes and macrophages express FcγRIIA, while FcγRIIB2 is expressed mainly on myeloid cells, although expression of FcγRIIB2 on monocytes dramatically varies between individuals [14]. In contrast to FcγRIIA expressed on pre-B cells, human FcγRIIB1 is expressed on all B lymphocyte subsets (pre-B and mature B cells). FcγRIIB expression by either $CD4^+$ or $CD8^+$ T cells is still controversial [27, 97] (Figure 4). In the experiments described below, both human activated lymphocytes and Jurkat cell line were lacking FcγRII expression (see section 5). FcγRIIC is found both on pre-B and mature B cells, monocytes and NK cells [21, 76].

FcγRIIB isoforms are unique among Fc receptors in exhibiting inhibitory properties. It is thought that FcγRIIC binding specificities are similar to those of FcγRIIB isoforms, but use signalling pathways similar to those of FcγRIIA. Contrary to the FcγRIIA and FcγRIIC isoforms, which both possess an ITAM within their intracytoplasmic tails, the FcγRIIB contains an immunoreceptor tyrosine-based inhibitory motif (ITIM) with the prototypic sequence motif I/L/V/S-x-Y-x(2)-L/V (see [116] for review). Indeed, the co-aggregation of RFcγIIB with activating receptors was demonstrated to negatively regulate cell activation triggered by all receptors containing ITAM domains (i.e. BCR, TCR, FcRs) both in vitro and in animal models [24, 55, 71, 121]. Human FcγRIIA favors phagocytosis, Ag presentation and calcium mobilization, while human and FcγRIIB molecules mediate endocytosis of their ligands and Ag presentation. While apoptosis is induced only after FcγRIIB aggregation, co-engagement of both BCR and FcγRIIB inhibit B cell proliferation and limit Ab secretion (see [92] for review).

Recombinant and viral MeV N binds in vitro to all FcγRII isoforms [91]. Immune complexes (physiological ligands) and MeV N (viral ligands) may share the same or overlapping binding sites on FcγRII, because the 2.4G2 mAb blocks both the murine FcγRII/immune complexes and FcγRII/MeV N interactions [91]. As mammalian FcγRIIs are conserved among species, MeV N is capable of interacting to both human and murine FcγRII expressed on APCs, including monocytes/macrophages, DCs and B lymphocytes, but not to T lymphocytes, epithelial cells and fibroblasts [67, 91].

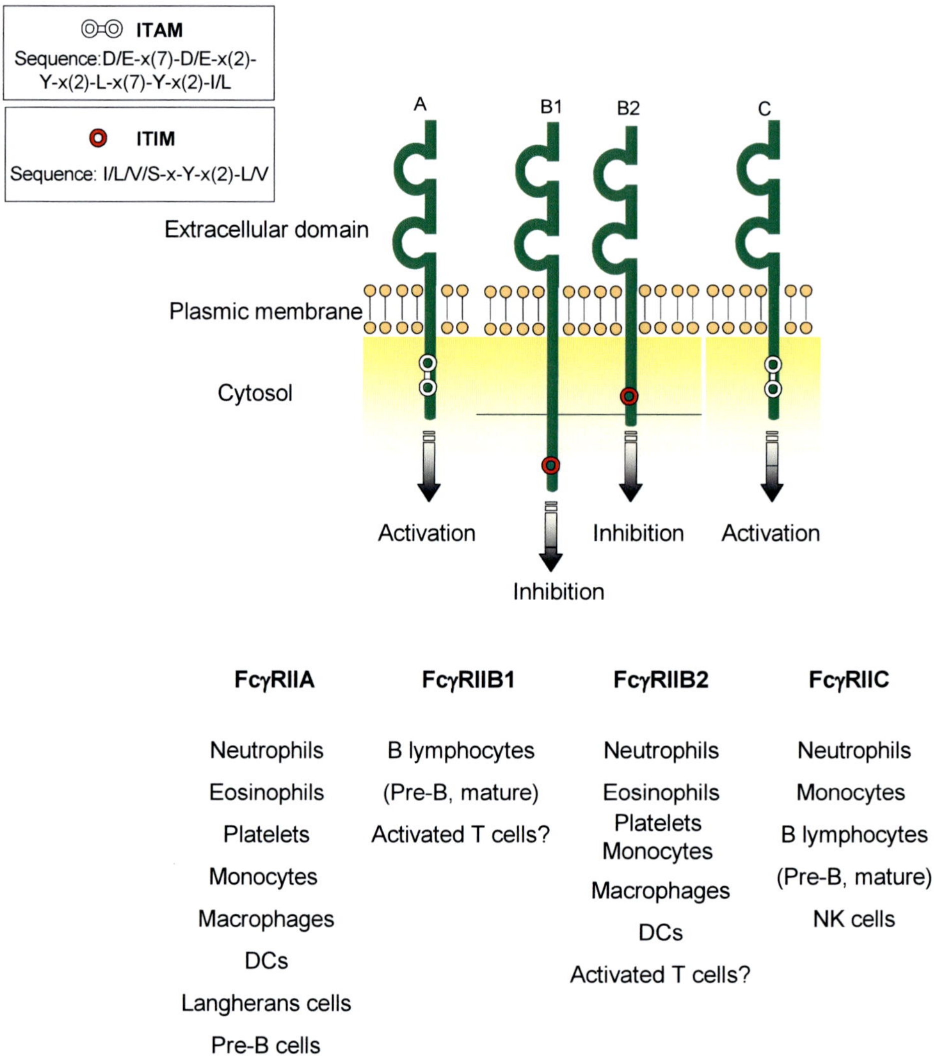

Figure 4. Structure and expression of FcγRII in humans. FcγRII are members of the Ig gene superfamily. FcγRIIs have been classified into FcγRIIA, FcγRIIB and FcγRIIC. A majority of FcγRII (FcγRIIA/C) possess the immunoreceptor tyrosine-based activation motif (ITAM) within the cytoplasmic domain. However, FcγRIIB has an immunoreceptor tyrosine-based inhibitory motif (ITIM). FcγRIIs are expressed on various types of hematopoietic cells involved in innate (polymorphonuclear cells, mast cells, NK, macrophages, monocoytes) and adaptive (DCs, T and B cells) immunity.

Morbillivirus N proteins are composed of a well-conserved hydrophobic and globular N_{CORE} domain, and a hypervariable disordered and hydrophylic N_{TAIL} domain (see Chapter 1 and Figure 2A). The region of MeV N responsible for binding to human FcγRIIB maps to N_{CORE} [66] (Figure 2C). Binding of NC-like particles to FcγRIIB is not a general property of any NCs. Although N from MeV, canine distemper (CDV) and peste des petits ruminants virus (PPRV) bind to FcγRIIB as detected by flow cytometry analysis, whereas N from rinderpest virus (RPV) doesn't [67]. However, RPV N binds to its host B cells, and with very low affinity to human FcγRIIB as detected using more sensitive techniques, such as surface plasmon resonance imaging [62]. The distinctive behaviour of RPV N may be due to its unique sequence property within the putative region of interaction with FcγRIIB. While Morbillivirus N proteins share 80 % homology over the entire N_{CORE} region, the homology drops to 40 % when considering the region spanning aa 122 to 144 [28]. This hyper-variable region, already described as an Agic region [38], is poor in hydrophobic clusters (see [59]) and may therefore form a loop exposed to the solvent. Within this loop, position 138 is occupied by a serine residue in all these Morbillivirus N proteins except for RPV N, where this serine is replaced by a glycine residue [28]. It is conceivable that binding to FcγRIIB may occur through this exposed region of N_{CORE} and that the serine to glycine substitution may lead to a conformational change by introducing a high degree of conformational freedom (Figure 2C). Moreover, MeV N binds to FcγRIIB more efficiently than does N_{CORE}. This observation may be accounted for the increased rigidity of N_{CORE} as compared to N ([70], see also Chapter 2). The lower flexibility of N_{CORE} may imply that conformational changes do not take place as efficiently as in the full-length MeV N, thus rendering sequential and/or conformational epitopes less accessible to FcγRIIB.

As MeV N can bind to both FcγRIIA/C and inhibitory FcγRIIB, we postulated that MeV N has a dual role in initiation and development of immune response. MeV N could activate cell expressing FcγRIIA or FcγRIIC (i.e neutrophils and NK cells respectively) and suppress activation of B cells via FcγRIIB. The effect on cells that express both activatory and inhibitory FcγRII remains to be determined.

3.3. Nucleoprotein Receptor (NR)

Previous studies focused on the characterization of FcγRII as MeV N receptor revealed MeV N binding to cells independent of BCR and FcγRII. Indeed, MeV N can bind to murine and human cell lines negative for the expression of N-specific BCR and FcγRII [67]. Moreover, this binding is not affected by blocking anti-FcγRII Abs. Conversely, the inhibition of MeV N binding to FcγRII by anti-FcγRII Abs is only partial (50 to 65% of inhibition) in cells expressing FcγRII [67]. These data support the existence of a novel cell surface receptor for MeV N, provisionally called nucleoprotein receptor (NR). MeV N binding to NR requires protein synthesis, is saturable and specific to MeV N, since non-related NC from rabies virus cannot bind to NR. Thus, NR is a protein that specifically binds MeV N [67]. NR is constitutively expressed on fibroblasts, epithelial, lymphoid and myeloid cells. NR is neither detected on resting T cells nor on red blood cells, but becomes expressed by human T cells when they are activated [67]. Constitutive expression of NR is not restricted to humans, since it is also detected on lymphoid and non-lymphoid cells from monkey and mouse (Figure 5).

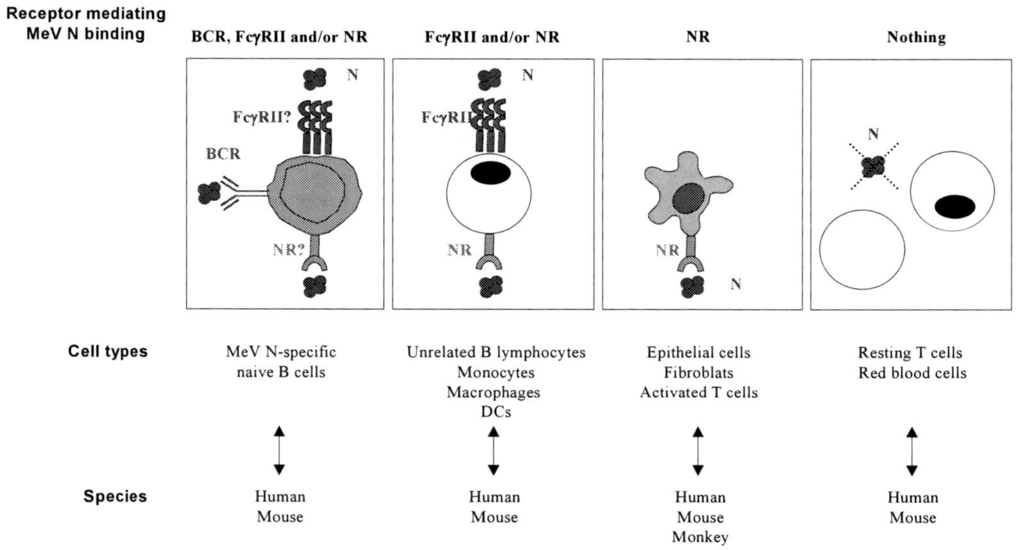

Figure 5. Interaction of MeV N with extracellular receptor(s) expressed by a large spectrum of cells from different species. Cells can be classified into 4 groups according to the expression of MeV N extracellular receptors. First, MeV-N specific B cells can bind N simultaneously via their high avidity N-specific BCR, and low avidity FcγRII and NR receptors. The second group includes cells that can bind N via both FcγRII and NR such as MeV N non-specific B lymphocytes, monocytes, macrophages and DCs. Cells expressing only NR belong to the third group, as fibroblasts, epithelial and activated T cells. The last group encompasses cells that do not bind to MeV N, such as resting T cells and red blood cells. Note that this cellular distribution is not restricted to humans but common to different species (from mouse to monkey).

NR detection on different cell species implies ubiquitous and conserved NR expression. According to this hypothesis, cell surface receptors of different species can equally bind to N proteins of *Morbillivirus* members that infect different hosts. Indeed, N from CDV, PPRV and RPV bind to both human and murine NR [67]. These results suggest that most *Morbillivirus* N proteins share a conserved NR on their respective hosts and that this receptor may derive from a common ancestral receptor. Alternatively, MeV N may bind to a group of heterogeneous NRs sharing similar binding properties. The biochemical characterization of NR is expected to provide new insights into the nature of this receptor.

The region involved in NR binding is localized within the hypervariable N-terminal region of MeV N, N_{TAIL} [67]. Specifically, NR binds Box1 (aa 401–420), which is well conserved among *Morbillivirus* members [28, 66] (Figure 2C). Box1 is also well conserved among wild-type and vaccine MeV strains, as judged by the comparison between the amino acid sequence of N_{TAIL} from Edmonston B, other vaccine strains and 48 wild-type strains [66]. Since three N proteins from other *Morbillivirus* tested so far can bind to NR, we looked for any conserved feature of the Box1 sequence. As shown in Figure 6, Box1 can be subdivided into smaller boxes defined by the sequence

[T]x[E/DD/ER/K]xx[R]xx[GPK/RQ]x[QV/ISF/TL] [28]. Synthetic peptides with the consensus sequence thus defined would help to further refine the NR binding motif on N_{TAIL}.

Although N_{TAIL} is highly sensitive to proteolysis [60], the N_{TAIL}/NR interaction remains possible in the context of an entire NC structure just after release in the extracellular compartment. Since MeV N binds to FcγRIIB and NR via N_{CORE} and N_{TAIL} respectively, MeV N can simultaneously bind to both FcγRIIB and NR on the same cell type in B cells, monocytes/macrophages and DCs. Therefore, we propose four groups of cells according to the set of MeV N receptors they expressed. B cells expressing the three MeV N receptors (i.e. N-specific BCR, FcγRII and NR) constitute the first group. The second group includes cells expressing both FcγRII and NR, such as monocytes, macrophages and DCs. Cells that express only NR belong to the third group. They include activated T cells, epithelial cells and fibroblasts. Finally, resting T cells and red blood cells belong to the group of cells lacking MeV N receptors, implying that MeV N cannot directly affect the biology of these cells via binding to cell surface receptors (Figure 5).

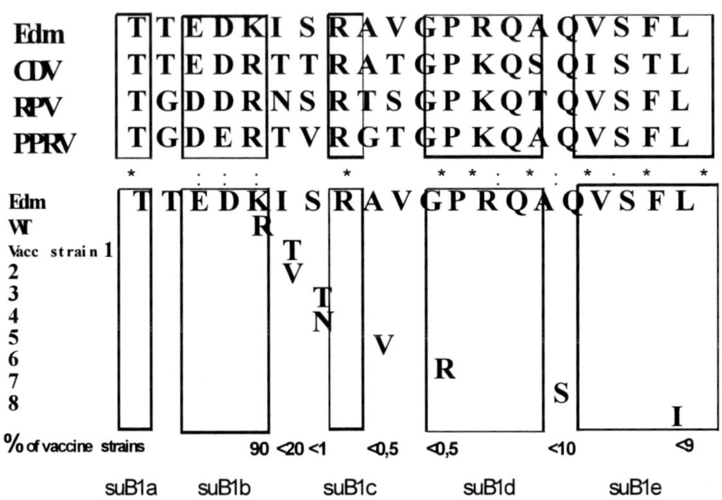

Figure 6. Sequences alignment between Box1-nucleoproteins of different Morbilliviruses: MeV, CDV, RPV and PPRV. (*) and (:) denote same or equivalent residues, respectively.

3.4. Do Viral and Intracellular Partners Interfere with Mev N Binding to Its Extracellular Partners?

When MeV N is released in the extracellular compartment after cell death [67], intracellular binding partners of N_{TAIL} are expected to be also released into the extracellular milieu (see Figure 2B and Chapters 3 and 4). The ability of the C-terminal domain X (XD) of MeV P [58] and of hsp72 [128], to compete out N_{TAIL} binding to NR was therefore investigated using purified proteins produced in E. coli [58, 127]. As a control, MeV N_{TAIL} was readily able to compete out with MeV N in a dose-dependent manner for binding to NR expressed on TEC cells (Figure 7A).

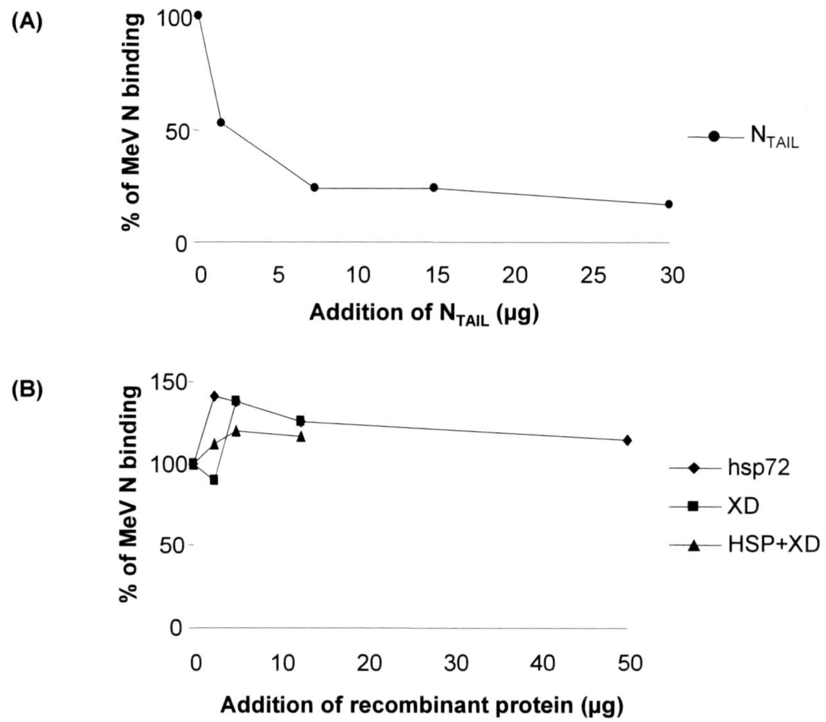

Figure 7. Simultaneous binding of MeV N to cell surface receptor NR and intracellular partners. (A) Competition of MeV N binding to NR by NTAIL. Binding experiments were performed by incubating MeV N (0.75 µg, corresponding to 50% of binding) with increasing amounts of NTAIL. MeV N binding was revealed by biotinylated anti-MeV N Cl120 mAb which recognizes the [133-140] N epitope. (B) Inability of XD and hsp72 to compete with NR for binding NTAIL. Competition binding experiments were performed by incubating NTAIL (1.25 µg, corresponding to 50% of binding) with increasing amounts of hsp72 or/and XD. NTAIL binding was revealed by biotinylated anti-MeV N Cl25 mAb which recognizes the [466-474] N epitope. The percentage of MeV N and NTAIL binding was calculated assuming that the mean of fluorescence intensity (MFI) observed with MeV N and NTAIL alone represents 100% of binding.

However, the addition of increasing amounts of hsp72 and/or XD had no effect on N_{TAIL} binding (Figure 7B). This is likely related to the fact that the NR binding site maps to Box1 (aa 401-420) [66, 67], which is far away from the XD and hsp72 binding sites which map to both Box2 (aa 489 to 506) and Box3 (aa 517 to 525) [15, 17, 58, 127]. There are many evidences for a strong association of N with P or hsp72, including their co-localisation in infected cells [115]. Thus, in the extracellular compartment N likely occurs as N/P and N/hsp72 complexes. The lack of competition between XD or hsp72 and N_{TAIL} for NR binding suggests that N bound to P or to hsp72 can bind to NR. This also suggests that the unstructured 125 aa long N_{TAIL} domain may be able to simultaneously accommodate several partners. However, this is not a general rule, as indicated by the observation that XD and hsp72 compete for binding to N_{TAIL} ([127], see also Chapter 3).

4. Induction of Specific Immune Response by MeV N

MeV infection is associated with N-specific humoral and cellular immune responses. In mice, immunization experiments performed with recombinant MeV N are strong enough to protect mice against lethal challenge with CDV, a Morbillivirus related to MeV [31]. These results suggest that in humans, the anti-N response contributes to viral clearance and long-term immunity. This section will focus on the current knowledge about the role of MeV N in the induction of efficient adaptive immune responses, before discussing the potential role of MeV N as a pathogen associated molecule pattern (PAMP) or an adjuvant.

4.1. Induction of $CD4^+$ and $CD8^+$ T Cells against MeV N

After infection, conventional DCs uptake and process Ags, then migrate from the primary site of infection to the lymph nodes and begin to present different Agic peptides on either MHC-class I or -class II to activate $CD8^+$ and $CD4^+$ T cells (known as T helper of T_H cells), respectively. During the acute phase of MeV infection, MHC-class I restricted CTL and, later on, MHC-class II restricted $CD4^+$ cells are detected in the blood. This $CD4^+$ T cell response occurs with a switch from T_H1 to T_H2 response (see below and [45] for a complete review). Accordingly, soluble CD8 and CD4 proteins are shed into the plasma, either transiently (CD8) or over 30 days (CD4) (Figure 1 and 8A). Soluble IL-2 receptor (sIL2R), a hallmark of activated T cells, is shed acutely at the onset of the rash and then slowly decreases over more than 100 days after the rash (Figure 8A). The $CD8^+$ and $CD4^+$ T cells specifically recognize the five major MeV proteins: N, P, M, F and H. Cell-mediated immunity against MeV N involves both MHC-class I-restricted $CD8^+$ T cells and MHC-class II-restricted $CD4^+$ T cells [54, 56, 111]. As infectious MeV enters the host cells by neutral pH fusion at the cell membrane, the incoming N protein is delivered into the cytosol of the APCs. There, it is readily available for cytosolic degradation and peptide loading onto MHC-class I resulting in efficient MHC-class I restricted presentation to $CD8^+$ T cells [19]. However, instead of fusing, some viral particles are endocytosed and delivered to the MHC-class II compartments, which results in classical MHC-class II presentation of N-derived peptides [36]. Furthermore, extracellular MeV N released by cytolysis could be internalized via the endocytic pathway resulting in presentation of MeV N peptides to MHC-class I and -class II in DCs. The MeV N specific T cell repertoire is still poorly characterized, even though it has been extensively studied in the early 90's. To our knowledge, all MeV N epitopes characterized so far are located within the N_{CORE} region ((Figure 2D), [8, 9, 25, 74]) and most of them are MHC-class I restricted (only 2 out of 21 MeV N epitopes are MHC-class II restricted) [25, 33, 54]. The strong $CD4^+$ response in the context of HLA-class II (HLA-DR1 restricted) against MeV N appears to contrast with the relatively low efficiency of Ag presentation of this protein compared to H [36]. Among the mechanisms that could occur, we mention the targeting of a soluble exogenous Ag to macrophages and DCs via their FcRs after its opsonization with specific Abs (see section 3.2) or direct MeV N binding to FcγRII expressed on DCs [67]. Both would enhance MHC-class II-restricted Ag presentation to $CD4^+$ T cells [1].

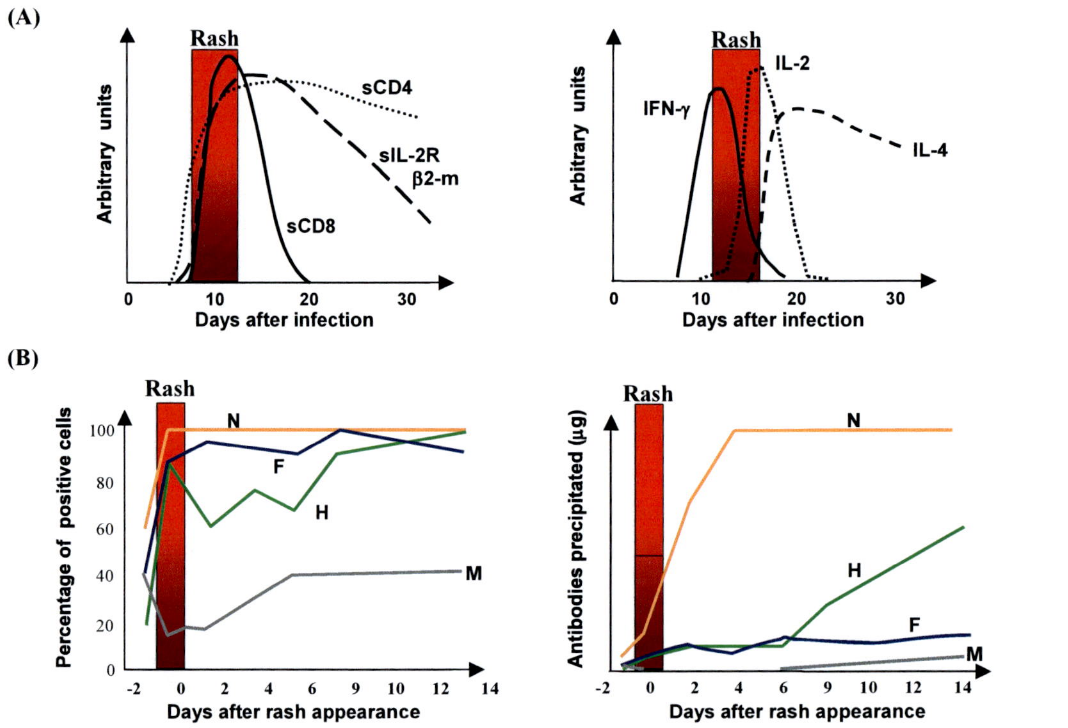

Figure 8. Evolution of MeV-specific immune response. (A) Schematic diagram changes in cell surface markers (left panel) and circulating cytokines (right panel) during MeV infection. Soluble IL-2 receptor, CD4 and CD8 are detected at the onset of the rash and then decline through time. The first cytokine detected in the plasma of infected patients is IFN-γ, which peaks at the beginning of the rash. The IL-2 peak is next detected, followed by IL-4 secretion. (B) Time course appearance of Abs to individual MeV proteins H, N, F and M at various times after appearance of the rash. Left panel represents the percent of viral protein expressing cells, while right panel indicates the relative amount of Abs precipitated. The first Abs produced in large amount are MeV N-specific [45].

However, it remains unknown whether receptor-mediated endocytosis via BCR, FcR, or any other MeV N-specific cellular receptor, plays a role in the presentation of N-derived epitopes to $CD4^+$ T_H cells and indirectly contributes to the massive synthesis of anti-N Abs.

4.2. High Production of Anti-N Antibodies

MeV humoral response is also detectable at the onset of the rash and rise rapidly thereafter (Figure 1 and 8B). Abs are induced to most MeV proteins with M protein being the less immunogenic one, since only few individuals make Abs to M. The earliest Abs produced are directed against MeV N and P. Neutralizing Abs against hemagglutinin (H) and fusion (F) glycoproteins are detected 2-4 days after the rash onset.

T_H2 cells are considered necessary for the full maturation of the humoral immune system. Indeed, T_H2 cytokines stimulate B-cell proliferation, Ab class switching, and Ab production. Interestingly, MeV mediates a shift towards a T_H2 response [5, 48, 78, 84] (Figure 8A). MeV-specific Ab response is initially IgM and IgA followed later by IgG, primarily IgG1. The most abundant and rapidly secreted Abs are against MeV N, and their absence is the most accurate hallmark of seronegativity. During the acute phase of MeV infection, Abs against MeV N are mainly IgG1, but also IgG3 and to a lesser extent IgG2, whereas MeV N-specific IgG4 and IgA responses are low or undetectable [29]. In some patients, low levels of N-specific IgM remained detectable a year after the onset of rash. Production of anti-N IgG Abs, and to a lesser extent anti-H, anti-F and anti-M Abs, is sustained long after clearing of the infection. Plasma IgE levels are reduced in children with measles [78], and there is no indication that specific IgE against MeV N or any other viral proteins are produced after natural MeV infection.

Since the Panum's epidemiological study in 1847, it is known that individuals can be protected against reinfection without re-exposure to MeV [87], but the mechanisms underlying the longevity of the humoral response are yet unknown. Follicular dendritic cells (FDCs) have been reported to retain Ag-Ab complexes (immune complexes) on their surface [107]. As Ab levels decline, FDCs could release more Ag to stimulate memory B cells and therefore repopulate declining plasma cell populations. At the onset of the rash, extensive replication of MeV coincides with the appearance of MeV N Abs. Thus, it is tempting to postulate that in lymphoid organs, specific Ab complexes are produced in large amounts and that N-containing immune complexes get trapped by FDCs. But can FDCs loaded for few days with MeV N immune complexes account for the decades of Ab production induced after infection? Another possibility is that Ab levels could be maintained by long-lived plasma cells without continued presence of persisting Ag (see [104] for review). Besides, memory B cells were found to constitutively express several toll-like receptors (TLRs) [11, 12]. When triggered by the corresponding TLR agonists, memory cells differentiate to plasma cells suggesting that microbial products can activate memory cells in an Ag-independent way for continuous plasma cell generation and homeostasis, thus contributing to maintain serum Ab levels (see [68] for a complete review). As MeV N Abs are the most abundant, one can hypothesize that MeV N triggers more Ag-specific plasma cells than the others viral proteins, and therefore induces more long-lived plasma cells and/or memory cells.

More investigations are required to evaluate the functional consequences of the rapid and abundant production of anti-N Abs. MeV N Abs can have only access to extracellular MeV N, which is either released in the extracellular milieu or expressed at the surface of live infected cells (see section 2). From an immunological point of view, MeV N Abs could limit MeV propagation by three different mechanisms. Firstly, MeV N Abs could bind to N expressed at the surface of infected cells and direct the cytotoxic activities of NK cells after cross-linking their FcRs, through a process called Ab-dependent cell-mediated cytotoxicity (ADCC). In MeV-infected individuals, the evolution of MeV-specific ADCC correlates temporally with the clearance of cell-associated virus [32]. Interestingly, the number and activation level of NK cells increase just after the onset of rash, when extensive MeV replication occurs and MeV anti-N Abs are secreted. Secondly, the binding of anti-N Abs to infected cells could also activate the complement, resulting in lysis of infected cells.

Interestingly, in vitro lysis of MeV-infected macrophages is induced after treatment with polyclonal rabbit Abs against MeV proteins, and is prevented after complement inactivation [41]. Finally, extracellular MeV N could be opsonized by MeV-specific Abs, targeted to FcRs expressed at the surface of APCs, and internalized by receptor-mediated endocytosis. Opsonized MeV N could then be degraded, and becomes accessible to the peptide-loading compartment, resulting in further enhancement of the MHC-class II-restricted T_H2 cell response that is required to boost the humoral responses.

Surprisingly, patients suffering from agammaglobulinemia (i.e. have little, if any, B cells) can control MeV infection, whereas patients with T-cell defects don't, indicating that an intact cellular, but not humoral, immunity is required to clear measles infection ([42], see [46] for review). Similarly, B cell depletion in rhesus monkeys does not significantly alter the course of MeV infection [89]. But, the lifelong immunity relies mainly on the presence of neutralizing Abs. Indeed, there is a strong correlation between levels of neutralizing Abs and protection, the best evidence being the protection of infants with maternal Abs. While the role of neutralizing Abs (i.e. against H and F proteins) in the protection against re-infection is widely documented (see [37] for review), the contribution of anti-N Abs is not established yet. These results imply that MeV Abs play a minor role in the control of MeV replication in the acute phase of the infection, thus limiting the role of neutralizing MeV Abs in protecting against a re-infection. The specific roles of humoral immune responses against MeV N during the primary infection and in the establishment of a lifelong immunity have still to be determined. The function of MeV N Abs could actually be subtler as revealed by Olszewska et al. [85]. The anti-N humoral responses could in fact decrease the incidence of complication following acute MeV infection. In a mouse model of MeV-induced encephalitis, MeV N immunization results in priming N-specific T and B cell responses, which confers moderate to complete protection following intracranial challenge with a neuro-adapted strain of MeV. Moreover, anti-N Abs can enhance the immunogenic properties of MeV N (see next section). Adversely, we propose that MeV N-specific Abs could also subvert the immune response and enhance N-mediated immunosuppression (see section 5).

4.3. Immunogenic Properties of MeV N

As previously discussed, the initial control of MeV infection requires an intact cellular immunity. $CD8^+$ CTLs play a crucial role in the control and clearance of MeV infection in the closely related rhesus monkey model [89, 90]. During the acute phase of the infection, MeV N stimulates a strong T-cell mediated cytotoxicity and an abundant production of specific Abs. This could relate to the peculiar immunogenicity of MeV N. In the absence of adjuvant, trans-epithelial injection of recombinant MeV N, or topical application on the buccal mucosa, is sufficient to induce N-specific cytotoxic $CD8^+$ CTLs and Abs [31, 69]. Immune responses triggered by injection or topical application of recombinant MeV N are strong enough to protect mice against lethal challenge with CDV, a *Morbillivirus* which shares with MeV the N281-289 Ld epitope ([31] and Figure 2D). It was also demonstrated that recombinant MeV N can adjuvant specific $CD8^+$ CTLs responses against poorly immunogenic Ags like ovalbumin [69]. This observation unravels the intrinsic immunostimulatory capacity of MeV N, and strongly suggests that during MeV infection, N-specific CD8$^+$ CTLs participate to viral clearance, recovery from disease and immunity against reinfection.

Ag capture, processing and presentation by DCs mostly determine the outcome of Ag-specific immune responses. *In vivo*, immunization by buccal application or transepithelial injection of recombinant MeV N was reported to induce a rapid and transient accumulation of MHC-class II$^+$/CD11c$^+$ cells with morphological features of conventional DCs [69]. MeV N was shown to mediate CCL20 chemokine induction, and the recruitment of circulating DC precursors that achieve differentiation in derma and epithelium. DC recruitment via CCL20/CCR6 interaction is required for subsequent cross-priming of CD8$^+$ CTLs after mucosal or skin immunization. Altogether, these observations indicate that MeV N stimulates first the recruitment of DCs, then their maturation and migration within draining lymph nodes to achieve T-cell cross-priming. The mechanism by which MeV N induces DC activation and/or maturation either directly via unknown receptor(s) or indirectly via the release of pro-inflammatory signals remains to be determined (see Chapter 4).

Table 1. Phenotype of monocyte-derived dendritic cells treated with MeV N

	None	MeV N alone[a]		Denatured MeV N[c]		MeV N +anti-N[d]		MeV N + anti-CD32[e]	
Markers		10 µg/ml	50 µg/ml	10 µg/ml	50 µg/ml	10 µg/ml	50 µg/ml	10 µg/ml	50 µg/ml
CD25	-	+/-	+	-	-	+/-	+	+/-	+
CD83	-	+	++	-	-	+	++	+	++
CD86[b]	-	+/-	++	-	+/-	++	+++	+/-	++

Immature dendritic cells (DCs) where derived from peripheral blood monocytes cultured for 6 days in the presence of GM-CSF and IL-4 as previously described (34). 5.10^5 immature DCs where incubated for 20 min at room temperature in the presence of indicated amounts of MeV N, then washed twice and cultured for 48 h in 0.5 ml of fresh culture medium before phenotypic analysis.

[a] -, +/-, +, ++ and +++ are used to gradually notify the percentage of cells (from 0 to 100%) expressing the marker analyzed by FACS.

[b] 100% of cells were positive for CD86 staining; in that case, -, +/-, +, ++ and +++ reflects the mean of fluorescence intensity (MFI) from low to high expression.

[c] MeV N was denatured at 90°C for 30 min.

[d] Culture was performed in the presence of 5 µg/µl of Cl25 anti-N mAb.

[e] DCs were pre-treated for 30 min in the presence of 10 µg/µl of anti-CD32/KB61 mAb before N was added. Data are representative of seven experiments performed on different donors.

In humans, MeV infection induces maturation of monocyte-derived DCs as assessed by phenotypic analysis [100]. However, MeV N alone was never tested in this system. To address the role of MeV N, we used immature DCs derived *in vitro* from purified human peripheral blood monocytes [34]. Cells where incubated in the presence of recombinant MeV N, then cultured *in vitro* before phenotypic analysis. The induction of CD25 and CD83 activation markers and the up-regulation of the CD86 co-stimulatory molecule demonstrate DC maturation after stimulation with recombinant MeV N (Table 1). This activity is dose-dependent and requires intact MeV N, as assessed by heat-denaturation experiments. This also demonstrates that maturation by MeV N is not carried by contaminant endotoxins that are known to be heat-resistant. Moreover, we show that DC maturation is enhanced by anti-N specific Abs. Accordingly, lower doses of recombinant MeV N can be used to yield similar effects (Table 1). Among other markers, maturation by MeV N is accompanied by the loss of the CCR6 receptor (data not shown).

This observation echoes to the studies performed in mice, where CCR6 down-regulation determines *in vivo* the migration of DCs from epithelium to draining lymphoid organs.

In general, DC maturation is regulated by specific receptors that bind elemental components of bacteria and viruses such as LPS or single-stranded RNA molecules [6]. Experiments performed both *in vivo* in murine models and *in vitro* in human cells, support the idea that NC-like structures formed by MeV N are recognized by DCs as PAMPs that signal danger. Interestingly, the ability of MeV to self-assemble into a NC-like structure and activate DCs is reminiscent of other PAMPs. For example, bacterial flagellin, a protein that assembles in a helical structure to form flagella, can activate immune response and DCs through specific binding to TLR-5 [50]. However, the receptor that mediates DC maturation in the presence of MeV N has not been identified yet. Incubation in the presence of the KB61 Ab, which blocks the interaction of MeV N with FcγRII, has no inhibitory effects on DC maturation, although it cannot be excluded that FcγRII contributes to the uptake of NCs (Table 1). This observation suggests that the maturation signal depends on a second receptor, such as NR [67] or an additional undetermined extracellular receptor. In addition, the region of MeV N responsible for the activation of DCs is not known. Two distinct domains within MeV N, N_{CORE} and N_{TAIL}, mediate self-assembly and interaction with NR, respectively [16]. We propose a model where the flexible N_{TAIL} domain, pointing at the surface of MeV NCs released in the extracellular compartment, could engage a receptor at the surface of DCs to induce maturation. However, final conclusions as to whether unstructured protein domains can effectively stimulate the immune response when exposed in the extracellular compartment and represent potential danger signals, awaits direct experimental evidence.

5. Induction of Immunosuppression by MeV N

As already discussed, an efficient immune response against MeV parallels a transient but profound immunosuppression (Figure 1C). This immunosuppression enhances the susceptibility towards other pathogens. Historically, MeV was the first pathogen recognized to cause immunosuppression. The only other virus known to induce such a profound suppression of T cells in humans is the immunodeficiency virus type I (HIV-I). In 1908, von Pirquet made a key observation: in patients suffering from measles disease, the tuberculin reaction disappears and this phenomenon lasts for several weeks after recovery [117]. Indeed, MeV induces a severe depression of cellular immunity, which accounts for the frequent additional infections with opportunistic pathogens causing to patients a high risk of severe courses of MeV. Many biological features of this immunosuppression have been characterized and include profound and prolonged abnormalities in cellular immune responses in infected hosts. At the onset of the rash, there is a severe lymphopenia affecting $CD8^+$, $CD4^+$ T and B lymphocytes [83] whereas the percentage of activated NK cells sharply rises. Typically, $CD4^+$ T lymphocyte levels lower than 200 cells/μl can be observed in young children with measles rash (Figure 1C). While $CD8^+$ T lymphocytes go back to normal level within one week, normalization of CD4 levels takes 2-3 weeks and that of B cells around 100 days [2, 84, 95] (Figure 1C). Surprisingly, the vast majority of these peripheral blood lymphocytes (PBLs) are uninfected because only a small number of cells (<0.01 %) get infected in patients. Therefore, indirect mechanisms rather than direct infection have to be involved to induce immunosuppression. One possible mechanism is that MeV infection interferes with the viability, maturation, and function of APCs, which may disrupt T cell-mediated immune responses [61, 101]. Additional non-exclusive mechanisms could imply viral proteins, since MeV H and F proteins at the surface of infected cells can bind to CD150, CD46 and/or an unknown receptor to induce a state of unresponsiveness to proliferation-activating signals in primary non-infected cells of the lymphoid and monocytic lineage (see [37] for review). In addition, binding of cell-free MeV to extracellular receptor(s) (NR and/or FcγRIIB) may affect the biology of non-infected cells. In this section, we will focus on the immunosuppressive properties of MeV and discuss the importance of MeV N in MeV-mediated immunosuppression.

5.1. Inhibition of IL-12 Synthesis and T Cell Proliferation by MeV N

The inability of lymphocytes isolated from patients with acute measles to proliferate in response to mitogenic and TCR stimulation *ex vivo* is a hallmark of MeV-induced immunosuppression. Several observations could account for this inhibition of T-cell proliferation: (i) MeV infection can block T cells in the G_0/G_1 phases of the cell cycle [99, 126], (ii) MeV can directly interact with lipid rafts leading to alterations of TCR-stimulated dynamics of signalling proteins within lipid rafts and deregulation of their activity [49] and (iii) direct infection of few monocytes and/or DCs could disrupt T cell mediated immune responses via impairment of cytokine secretion [101].

A defect in IL-12 secretion was reported in measles patients since plasma IL-12 levels were found to be within the lower limit of normal range [78, 84]. When stimulated *ex vivo*, PBMCs from MeV infected patients display reduced proliferation, poor IL-12 secretion and recovery to normal values takes more than 32 weeks [3, 47, 124]. Impairment of DC functions is characteristic of measles and includes impaired priming of naive T cells because of defective IL-12 production [34]. Amongst the mechanisms that have been described to inhibit IL-12 secretion, MeV N was shown to inhibit IL-12 production by human CD40-activated DCs *in vitro* [100]. FcγR-dependent mechanisms of suppression of IL-12 synthesis have been demonstrated in human monocytes/macrophages [44]. Therefore MeV N binding to FcγRIIB on DCs may represent a key factor for IL-12 inhibition in MeV-infected patients [119]. This question was addressed in a murine model where recombinant MeV N suppresses both $CD4^+$ and $CD8^+$ T-cell dependant DTH and hypersensitivity responses (HSR) to a third party Ag [62, 72]. Similarly, CDV and PPRV N-mediated reduction of inflammatory reactions are caused by weakening DC functions such as IL-12 secretion, that abolishes specific $CD8^+$ T cell proliferation. However, RPV N neither binds efficiently to FcγRIIB (see section 3.2) nor does exert any immunosuppressive effect in mice. This inhibition requires FcγRIIB because it is no longer observed in FcR null mice [62, 72]. Comparison with highly homologous *Morbillivirus* N proteins shows that the ability of the latter to dampen the HSR and to exert an immunosuppressive effect in mice correlates with their ability to bind efficiently to FcγRIIB ([62] and section 3.2). Thus, N-mediated engagement of FcγRIIB likely participates to the inhibition of DC functions, including suppression of IL-12 secretion associated with the T_H2 polarization observed during MeV infection.

Table 2. Both FcγRII and NR contribute to the suppression of cellular proliferation after binding MeV N

Cell types	MeV N binding (MFI)[a]	FcγRII expression[b]	Predictive NR expression[c]	Inhibition of cell proliferation (%)[d]
Normal cells:				
Human epithelial cell (TEC)	+++	-	+	*87*
Murine fibroblast cell line				
L	+	-	+	*47*
L HuFcγRII (L-CD32)	++	+++	+	*52*
Human CD3/CD28 activated T cells	+	-	+	*100*
Tumor cells:				
Jurkat (Human acute T cell leukemia)	+	-	+	*61*
Human burkitt B lymphoma				
THOMAS	++	+	ND[e]	*39*
JiJoye	++	+	ND	*30*
Murine B lymphoma				
IIA1.6	++	-	+	*25*
IIA1.6 mFcγRIIB1	+++	++	+	*70*
IIA1.6 mFcγRIIB1Δcyt(II IC5)	+++	++	+	*33*
Human metastatic melanoma (Me 275)	+	-	+	*22*
Human myeloma (KMM1)	+++	-	+	*65*

[a] Cells were incubated with 50 μg/ml of purified recombinant MeV N from insect cells and binding was revealed with biotinylated anti-MeV N (Cl25) mAb, followed by streptavidin-PE as described by (67). MFI of MeV N binding is indicated as follows: - no binding; + MFI between 10 and 50; ++ MFI between 50 and 100; +++ MFI>100. All labelings were analyzed by flow cytometry, and data are the mean of three separate experiments (SD values are < 15%). NA, not applicable.

[b] FcγRII expression was determined using anti-human or -murine FcγRII mAbs and indicated as: no expression (-) or low, intermediate or high FcγRII expression (+, ++, and +++ respectively).

[c] Putative second receptor expression was deduced by analysis of MeV N binding when FcγRII expression is lacking.

[d] Spontaneous and activated cell proliferation after MeV N binding or FcγRII aggregation is indicated. Cells were cultured with 100 μg/ml of MeV N for 2 days. For inhibitory proliferation assays, the CPM value obtained in the absence of MeV N was considered as 0% of inhibition.

[e] ND, not determined.

In addition to the modulation of cytokine production by DCs, T cell proliferation can be directly suppressed by MeV N. Although human T cells freshly purified from the peripheral blood do not constitutively express NR or FcγR, NR is induced after T cell activation by either phorbol myristate acetate plus ionomycin or Abs to TCR (CD3) and co-stimulatory molecules (CD28) (Table 2 and [67]).

Interestingly, auto-proliferative T cells, such as Jurkat, constitutively express NR strongly suggesting that NR is a marker of T cell activation (Table 2). NR engagement by MeV N dramatically suppresses both mitogen-induced and spontaneous T cell proliferation in a dose-dependent manner (inhibition varies from 50% with 25 μg/ml of MeV N to 100% with 100 μg/ml of MeV N). MeV N_{TAIL}/NR interaction blocks T cells in the G_0/G_1 phases of the cell cycle rather than inducing apoptosis, as discussed later (see section 5.3). Thus, NR is an inhibitory receptor and MeV N directly inhibits the proliferation of non-infected T cells *in vivo.*

In summary, T cell proliferation is inhibited and/or modulated by MeV N either directly by binding to NR expressed on activated T cells or indirectly by cross-linking FcγRIIB on DCs, which dampens IL-12 secretion and accounts for the T_H1 to T_H2 switch (Figure 9A).

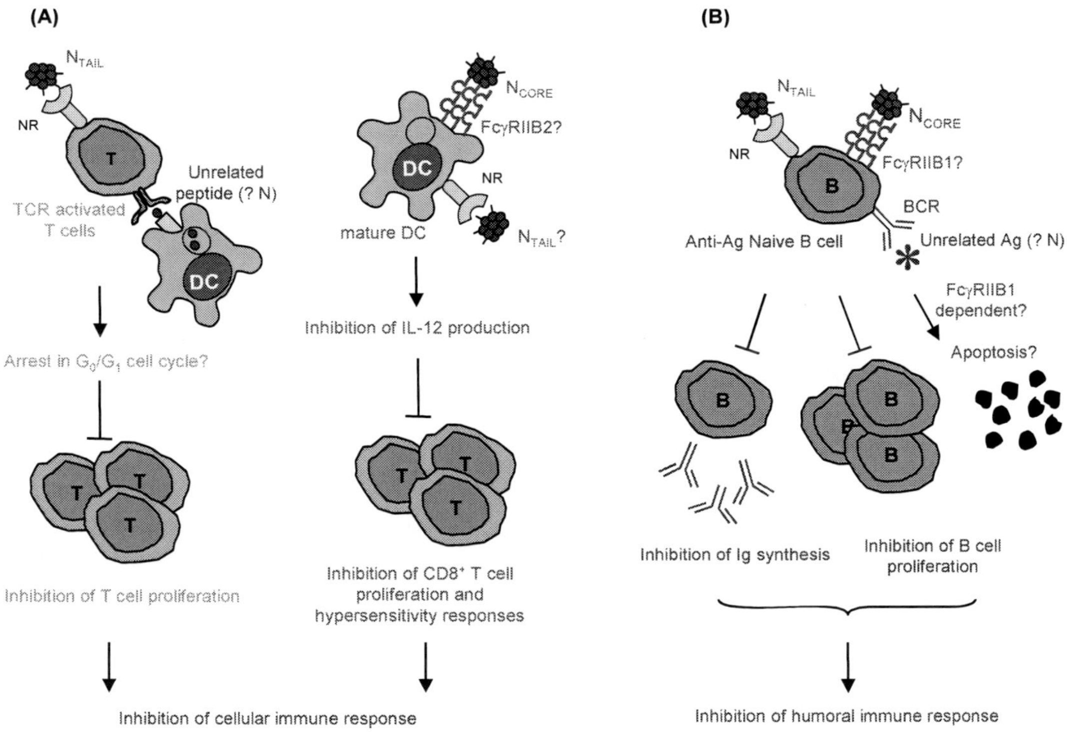

Figure 9. Potential consequences of MeV N interaction with FcγRII and NR in immunosuppression. (A) MeV N can alter DCs/T lymphocytes by two distinct mechanisms. First, NTAIL can bind to NR expressed by activated T cells to suppress their activation and subsequent proliferation. Secondly, NCORE can directly interact with both FcγRII and NR on DCs to inhibit their function, including IL-12 secretion, thus precluding efficient activation of T lymphocytes. (B) MeV can simultaneously bind to FcγRII and NR on B cells via NCORE and NTAIL, respectively. MeV N binding to inhibitory receptors can inhibit both cell proliferation and Ab secretion by activated B cells. It is still unclear whether MeV N-FcγRII cross-linking triggers B cell apoptosis.

5.2. Inhibition of B Cell Activities by MeV N

In MeV-infected patients, immunosuppression also affects the B cell compartment with a B cell lymphopenia, inhibition of mitogen-induced B cell proliferation [30, 118] and impairment of Ab response to typhoid vaccination [123]. *In vitro* data also indicate that MeV infection inhibits B cell proliferation and subsequent Ab production [23, 123]. Contrary to T cells, which express only NR, B cells simultaneously express BCR, and two inhibitory receptors for MeV N: FcγRIIB1 and NR. Below, we will discuss the potential role of N in MeV-induced inhibition of B cell functions.

5.2.1. MeV N-Mediated Inhibition of B Cell Proliferation

To date, no inhibition of mitogen-induced B lymphocyte proliferation has been ascribed to MeV N. However, we show here that recombinant MeV N inhibits spontaneous human B lymphoma proliferation (Table 2). Strong MeV N binding was observed on all cell lines tested. MeV N binds to burkitt B lymphoma cell lines (THOMAS and JiJoye) *via* FcγRII (Table 2). Expression of NR on these cells remains to be determined. These two cell lines display reduced proliferation in the presence of MeV N (up to 39% inhibition, Table 2). To distinguish the role of FcγRIIB1 and NR in this inhibition, we used the murine B lymphoma cell line (IIA1.6 not expressing FcγR) transfected or not with expression vectors encoding intact murine FcγRIIB1 (IIA1.6 mFcγRIIB1) or FcγRIIB1 lacking the intra-cytoplasmic tail of the receptor (IIA1.6 mFcγRIIB1Δcyt, II IC5, [1]). As illustrated in Table 2, MeV N binds to these three cell lines, with the highest binding efficiency for cells expressing mFcγRIIB1. Similar binding activities are observed with IIA1.6 mFcγRIIB1 and II IC5, indicating that deletion of the FcγRIIB1cytoplasmic tail does not affect the binding of MeV N to this receptor. These results also indicate that MeV N simultaneously binds to both murine FcγRIIB1 and NR. As expected, MeV N suppresses spontaneous proliferation of IIA1.6, IIA1.6 mFcγRIIB1 and II IC5 (Table 2). 25 to 33% inhibition of IIA1.6 and II IC5 cell proliferation was observed, showing that a fraction MeV N inhibitory activity relies on NR, and not on FcγRIIB1 intra-cytoplasmic tail. Distinctively, MeV N induces 70% inhibition of IIA1.6 mFcγRIIB1 cell proliferation, indicating that mFcγRIIB1 is also involved in suppression of cell proliferation (Table 2). It is very unlikely that MeV N renders B cells anergic, since IIA1.6 mutants can resume their proliferation when MeV N is removed from the milieu (not shown).

As the function of specific MeV N Abs produced following measles infection has not been fully unravelled (see section 4), we examined the role of MeV N-specific Abs in our model of MeV N-mediated inhibition of murine B lymphoma cell proliferation. Addition of anti-N Cl25 mAb did not significantly affect MeV N-induced inhibition of IIA1.6 proliferation nor N-induced inhibition of IIA1.6 mFcγRIIB1 proliferation at normal doses (100 μg/ml) (Table 3 and data not shown). This result indicates that MeV N-specific Abs do not sequester MeV N to hamper its interaction with FcγRIIB1. To our surprise, anti-N Cl25 mAb dramatically enhanced N-mediated inhibition of IIA1.6 mFcγRIIB1 proliferation when using very low N concentration (0.05 μg/ml, Table 3). This effect was (i) specific of anti-N Cl25 mAb, since low doses of MeV N alone did not have any effect on IIA1.6 mFcγRIIB1 proliferation and (ii) strictly dependent on mFcγRIIB1 expression because II IC5 cell proliferation was not suppressed by MeV N and anti-N Cl25 mAb (Table 3). It is puzzling that Abs to MeV N enhance, rather than protect, the inhibitory effect on cellular proliferation. This finding raises the point of an immunosuppressive property of anti-N Abs secreted after natural measles infection. As the beginning of the immunosuppression coincides with the production of anti-N Abs, it is tempting to speculate that these Abs are a key factor in MeV-induced B cell lymphopenia as they favor MeV N antiproliferative properties upon cross-linking of a small amount of MeV N to FcγRIIB.

Table 3. Effect of MeV N-mAb on N-mediated inhibition of cellular proliferation of IIA1.6 cell lines

Cell types[a]	Inhibition (%) of cell proliferation after the addition of:			
	N (100 μg/ml)[b]	N (0.05 μg/ml)[c]	N (0.05 μg/ml) + anti-N mAb[d]	Anti- FcγRII mAb[e]
IIA1.6	25	0	0	0
IIA1.6 mFcγRIIb1	70	0	55	58
IIA1.6 mFcγRIIb1 Δcyt (II IC5)	33	0	0	0

[a] IIA1.6 lack FcγRII, but express NR. IIA1.6 mFcγRIIb1 express both NR and mFcγRIIb1, while II IC5 express NR plus a truncated form of mFcγRIIb1 lacking its intracytoplasmic tail. Results are expressed as percentage of inhibition after 2 days of culture. The CPM value obtained in the absence of MeV N was considered as 0% of inhibition.

Cells were cultured with:

[b] 100 μg/ml of MeV N as in Table 2.

[c] 0.05 μg/ml of MeV N.

[d] 0.05 μg/ml of MeV N in the presence of 5 μg/ml anti-MeV N (Cl25) mAb.

[e] 5 μg/ml of anti- mFcγRIIb (2.4G2) mAb to aggregate FcγRII.

Note that MeV N was purified from insect cells. Data are expressed as the mean of three separate experiments (SD values are < 15%). For murine B lymphoma cell lines, IIA1.6 expressing or not FcγRII were treated for 2 days with either murine anti-FcγRII mAbs (4 μg/ml) only, or 100 μg/ml of MeV N only, or 0.05 μg/ml of MeV N in the presence of murine anti-FcγRII mAbs (4 μg/ml).

5.2.2. MeV N-Mediated Inhibition of Antibody Production

B cell differentiation into plasma cells requires a proliferation phase prior to the Ig secretion. As MeV N can down regulate B cell proliferation, the role of MeV N in the inhibition of Ab production will now be discussed. Another classical hallmarks of MeV-induced immunosuppression are the lack of responsiveness of non-infected PBLs isolated from patients in response to Ags or mitogens. Specifically, MeV infection inhibits Ig secretion from pokeweed mitogen-stimulated PBL cultures and impairs the proliferative response of B cells to mitogens [23, 123]. There are several hypotheses to explain the MeV-induced Ig suppression, including direct infection of B cells [75]. However, the fact that only a minor fraction of PBMC is actually infected *in vivo* suggests again that indirect mechanisms are involved. Interestingly, extracellular MeV N was demonstrated to dramatically inhibit (with inhibition values ranging between 49 and 62%) polyclonal IgM, IgG, and IgA secretion from purified tonsillar B cells costimulated with IL-2, IL-10 and either immobilized anti-Ig Abs or CD40L [91]. The MeV N-mediated IgG inhibition is to a certain extent very similar to the cross-linking of FcγRIIB induced by anti-FcγRII Abs [53]. Therefore, we propose that MeV N binding to FcγRII down modulates the capacity of B cells to differentiate into Ig-secreting cells. Inhibition of Ig synthesis by MeV N is independent from BCR, as MeV N down regulates the Ig response when B cells get activated by CD40 cross-linking. One plausible explanation is that the sole cross-linking of FcγRIIB is sufficient to transduce a negative regulatory signal. FcγRIIB inhibitory activity has also been reported after aggregation of this receptor independently from BCR signalling. In this case, an inhibitory signal is generated through Bruton's tyrosine kinase (Btk) and Jun kinase (Jnk), independently of the ITIM and the SH2 containing inositol 5-phosphatase (SHIP) and induces B cell apoptosis rather than an inhibition of cell proliferation [13, 86, 88]. This hypothesis is compatible with the fact that both native and recombinant MeV N have an oligomeric structure thus susceptible to efficiently cross-link FcγRII on B cells. It is noteworthy that no inhibition of B cell proliferation was observed in the presence of low amounts of MeV N ([91], Table 2 and data not shown). In agreement with previous reports, Ag-Ab complexes bound to murine B lymphocytes via FcγR can inhibit the Ab response of the latter, but not their proliferation in response to $F(ab)'_2$ anti-μ and lymphokines, suggesting uncoupled events between B cell proliferation and Ig secretion [20, 110]. Further investigations are needed to determine the molecular pathway(s) involved in MeV N-mediated inhibition of Ab-secretion and potential role of NR.

In conclusion, both FcγRIIB1 and NR contribute to the inhibition of cellular proliferation of B cells upon MeV N binding and the subsequent Ab production. Can MeV N impair induction of normal B cell responses *in vivo* as well? This is likely since cross-linking of FcγRIIB1 by Abs greatly reduces the generation of B cell response *in vivo* (see [51, 92] for review and Figure 9B).

5.3. MeV N-Mediated Inhibition of Non-Lymphoid Cell Proliferation

Contrary to FcγRIIB, NR expression is not restricted to immune cells (see section 3). In this section, we address the point as to whether MeV N could bind to normal and tumoral non-lymphoid cells and modulate their biology. Specifically, we will discuss the structural and functional consequences of MeV N release in secondary lymphoid organs, as illustrated by the thymic epithelial cell (TEC) model. Finally, we will discuss whether the MeV N anti-proliferative properties could be exploited to suppress tumour growth.

5.3.1. Inhibition of Normal Non-Lymphoid Cell Proliferation by Mev N: A Bystander Effect on T Cell Proliferation?

The global antiproliferative action of MeV N has been first illustrated in the TEC model. The thymus is a primary lymphoid organ that supports thymocyte survival and differentiation into mature T cells. Briefly, thymic microenvironments provide survival, proliferation, and differentiation signals for thymocytes development (see [65] for review). Within the cortex, positive selection of immature T cells expressing functional TCR is mediated mainly by cortical TEC, whereas in the medulla, negative selection of developing autoreactive T cells is ensured by APCs of hematopoietic origin and, to a lower extent, the TEC subset. Autopsies of patients with fatal MeV infection demonstrated lesions in the thymus with degenerative and/or necrotic changes associated to a rapid and predominant loss of the thymic cortex [22, 109, 122]. A marked involution of the thymic medulla was also observed, and MeV Ags were found in Hassall's corpuscles and in adjacent MeV-infected thymic stromal cells [22, 64, 77, 96, 102, 122, 125], indicating that the thymus is susceptible to MeV infection. This was confirmed in a SCID-hu mouse model, where thymic stromal cells are targeted by MeV after direct intrathymic inoculation [4].

Viruses that spread to the thymus mediate thymocyte deletion through two distinct mechanisms: replication within immature T cells leading to their apoptosis and replication within thymic stroma cells leading to the disruption of thymic micro-environment [4, 40]. Because of MeV replication, infected TECs form massive syncytia. The presence of large syncytial multinucleated giant cells, the so-called WFCs, has been described in the thymus, as well as in other lymphoid tissues, during the early phase of measles infection [106]. The syncytia eventually die by apoptosis, releasing MeV N in the extracellular milieu (refer to section 2). Extracellular MeV N released after apoptosis of MeV-infected cells can bind to uninfected neighbouring TECs to inhibit their proliferation [67]. Although TECs do not express FcγR, MeV N binding to NR expressed by TECs leads to signal transduction with modulation of Ca^{2+} influx. Indeed, MeV N alone induces a small, but significant, increase in the intracellular free Ca^{2+} basal level. Moreover, co-stimulation with ionomycin and MeV N elicits a sustained Ca^{2+} signal. As changes in the Ca^{2+} concentration affect the balance between proliferation and differentiation in epithelial cells, it is not surprising that MeV N, especially N_{TAIL}, inhibits TEC proliferation in a dose dependent manner by arresting cells at the G_0/G_1 phases of the cell cycle (Table 2) [67]. However, cell-free MeV N does not induce TEC differentiation, or IFN-α/β production or apoptosis (Valentin, H., unpublished data). In this context, MeV N mediated growth arrest of uninfected TECs would affect immature T cell survival by signal deprivation, particularly in infants, in whom ontogenesis of the immune system requires functional thymic micro-environment. Further investigations are required to sustain this hypothesis. We can speculate that the *in vitro* destruction of thymic epithelium caused by MeV N may deliver abnormal signals to thymocytes and/or affects both positive and negative selection of thymocytes.

5.3.2. Inhibition of Tumoral Non-Lymphoid Cell Proliferation by MeV N

The inhibitory properties of MeV N are not restricted to epithelial cells. As illustrated in Table 2, MeV N binds to L cells (murine fibroblasts), with the highest binding ability being observed for L cells expressing mFcγRIIB1 (L-CD32) (Table 2, not shown). The spontaneous proliferation of L cells and L-CD32 cells is also impaired by MeV N (47% to 52%). These results are consistent with the observation that MeV N anti-proliferative properties are mediated by both FcγRII and NR binding. These results also point out that MeV N anti-proliferative properties are not restricted to cells of the lymphoid lineage and thus could play an important role in the MeV-induced physiopathology.

As MeV N can efficiently suppress the mitogen- and spontaneous- proliferation of a broad spectrum of normal cells, we next investigated the consequence of MeV N binding on tumour cell proliferation. We used two human metastatic melanoma (Me 275 and HT144) and myeloma (KMM1) cell lines that do not express human FcγRII (Table 2), as well as HT144 cells transfected with FcγRIIB (HT144IIB) expressing human FcγRII [66]. As expected, both HT144 and HT144IIB cells efficiently bind MeV N, indicating that MeV N simultaneously binds to both human FcγRIIB and NR (Table 2 and [66]. In addition, MeV N impairs proliferation of tumour cells by 22 to 100% (Table 2 and [66]). These results suggest that MeV N has global anti-proliferative properties mediated by FcγRII and NR surface receptors.

In this model, the N_{TAIL}/NR interaction induces an arrest in the G_0/G_1 phases of the cell cycle and subsequently a decrease in the percentage of cells in both S and G_2/M phases of the cell cycle [66]. Using N_{TAIL} deletion proteins, we reported that the 401-420 region of MeV N is required for binding to NR, and thereby is responsible for cell proliferation inhibition. Interestingly, this region (residues 401-420) is well conserved among Morbillivirus members [28] and among wild-type and vaccine MeV strains, as judged by the comparison between the amino acid sequence of N_{TAIL} from Edmonston B, other vaccine strains and 48 wild-type strains (see Figure 6). Similarly to N_{TAIL}, N_{CORE} binding to FcγRIIB also elicits an arrest in the G_0/G_1 phases of the cell cycle, but only the N_{CORE}/FcγRIIB interaction eventually triggers apoptosis through caspase-3 activation in HT144IIB cells [66]. As cross-linking of FcγRIIB by mAbs triggers cell apoptosis through Btk and Jnk [13, 86, 88], we postulated that the N_{CORE}/FcγRIIB interaction recruits Btk and Jnk to induce apoptosis. However, the link between inhibition of cell proliferation and induction of apoptosis mediated by both MeV N and N_{CORE} after binding to FcγRIIB1 is difficult to establish, and we cannot exclude the possibility that the induction of apoptosis is the consequence of suppression of cell proliferation.

Altogether our results indicate that MeV N can interact with lymphoid cells via both FcγRIIB and NR to efficiently prevent cellular proliferation by inducing an arrest in the G_0/G_1 phases of the cell cycle and apoptosis. Alternatively, both lymphoid and non-lymphoid cells, which only express NR, would be blocked in the G_0/G_1 phases of the cell cycle upon MeV N binding to NR. Besides, the global and strong antiproliferative MeV properties could potentially be used in tumor therapies as N can regulate the proliferation and/or apoptosis of malignant cells according to the receptor used.

6. Model for the Interaction of MeV N with Receptors: How Induction of Immune Response and Immunosuppression Cohabit?

We propose a model of the paradoxical immunostimulating and immunosuppressing effects of MeV N during MeV infection (Figure 1). Significant amounts of extracellular N may affect the balance between induction versus inhibition of the immune response. Indeed, the immunostimulatory activities of N are observed using low doses of MeV N [31, 85], whereas high doses are required to suppress the immune response [66, 67, 72, 91, 101].

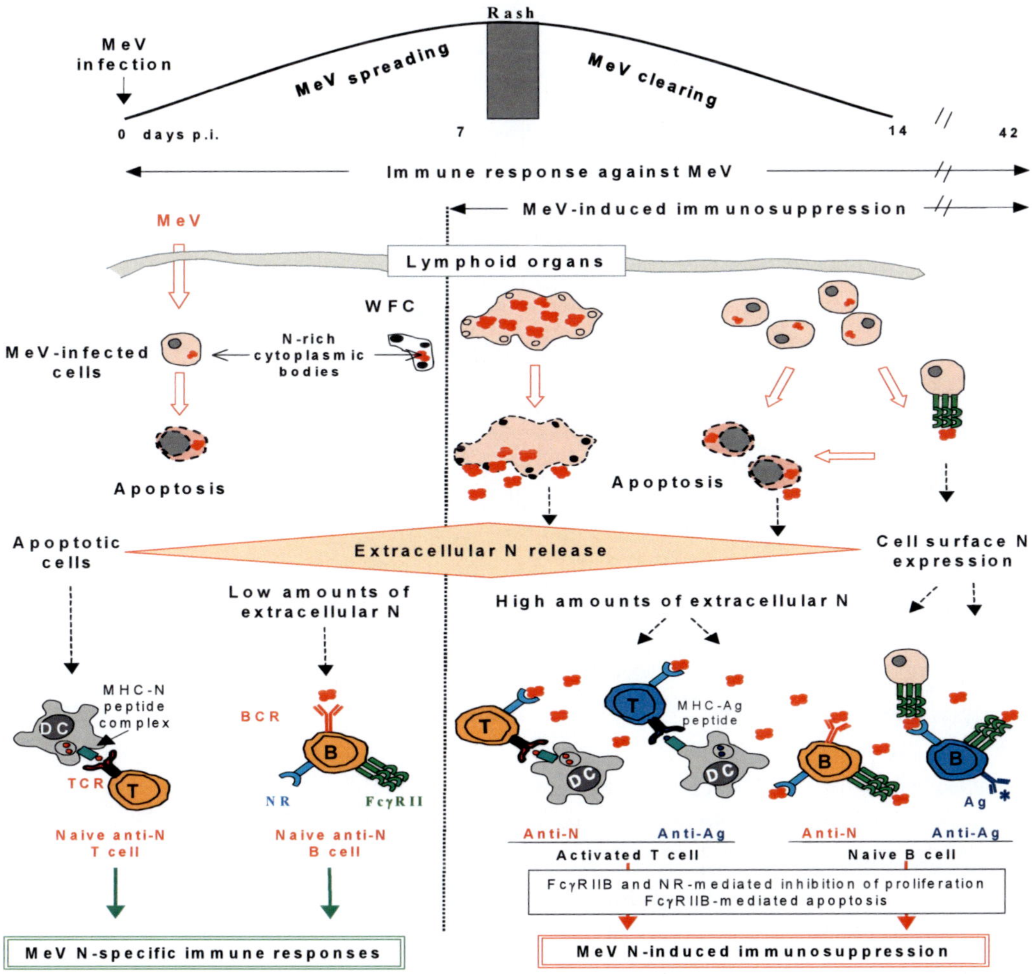

Figure 10. A two-step model of induction of early MeV N specific immune response and late generalized immunosuppression by N protein. During the course of MeV infection, MeV N presents paradoxical properties modulating the immune response both in a positive and a negative way. The model predicts that available increasing amounts of extracellular MeV N govern these two opposite activities. At the early stage of infection (left), few MeV particles spread and replicate within the secondary lymphoid organs. This results in the apoptosis of infected cells and in the generation of few giant multinucleated cells (WFCs), containing numerous intracytoplasmic inclusion bodies rich in N protein. MeV infected cells can express viral peptides on their MHC class I molecules, that will activate cytotoxic cells (NK, DCs or $CD8^+$ T cells) and lyse the infected cells. As a result, cytosolic N is released intro the extracellular compartment. Apoptotic bodies derived from infected cells are processed by DCs thus leading to efficient activation of MeV specific MHC restricted $CD8^+$ and $CD4^+$ T cells. Low amounts of MeV N are also released in the extracellular milieu and bind to specific and high avidity B cell receptors (BCR) resulting in efficient activation of N-specific B cells. Alternatively, N-specific Abs could enhance the FcγRII-mediated inhibition of proliferation observed with very low amounts of MeV N. At the rash onset (right), intensive MeV replication occurs in the secondary lymphoid organs with massive cell death of infected cells leading to the accumulation of high amounts of intracellular and extracellular N. Cell-free MeV N can bind to low-avidity FcγRII and NR receptors and disrupt MeV non-specific B- and T-cell-mediated responses, while N-specific effector cells are sustained.

MeV initially replicates in the respiratory tract and then spreads to local lymphoid tissue, where virus replication can occur in macrophages and/or DCs. Upon entry of MeV into the lymphoid organs via MeV-infected DCs, non infected DCs could engulf apoptotic infected cells, then display N-derived peptides in complex with MHC class-II molecules, and activate N-specific helper $CD4^+$ T cells (Figure 10).

MeV N is synthesized within the cytosol of MeV-infected cells, before being released in the extracellular milieu after death of infected cells and/or translocated to the plasma membranes of infected cells via FcγRII. Other sources of cell-free MeV N can be giant multinucleated WFCs [106] observed both in the germinal centers and in the follicular areas of the draining lymph nodes [80], lysis of MeV-infected cells by NK cells, complement and/or CTLs (see section 3). Therefore, low amounts of extracellular N would become directly available to naive immune cells and could bind to B lymphocytes expressing high avidity N-specific BCR (81), and prime them for $CD4^+$ T cell dependent activation [43, 79] (Figure 10). Meanwhile, the secondary viremia allows the spreading of the virus to other lymphoid organs, including the thymus and may alter the generation of functionally T cells, as previously discussed in section 4. Later on, intensive replication occurs with induction of WFCs and cell apoptosis leading to the release of high amounts of N in the extracellular milieu. The newly synthesized anti-N Abs could enhance the binding of MeV N to FcγRIIB to inhibit B and DC functions. Then, the level of extracellular MeV N reaches a sufficient level for efficient binding to its low avidity receptors FcγRIIB and NR [91] (Figure 10). Since MeV N is a potent inhibitor of cell proliferation through additive inhibitory effects of FcγRIIB and NR or efficient cell-cycle arrest through NR [66, 67, 105], MeV N could trigger inhibitory signals, which will prevent non-specific B and T cell proliferation (Figure 10). This inhibitory effect may be enhanced by the anti-N antibodies cross-linking to FcγRIIB.

In conclusion, exogenous cell-free MeV N likely contributes to the panlymphopenia observed during acute MV infection and the disruption of both B- and T-cell-mediated responses. As MeV prevents the acquisition of lymphocyte functions, but does not inhibit these functions when established [35], MeV N would be unable to down-regulate the virus-specific immune effector functions, thus explaining how an efficient immune response against MeV can be elicited despite the strong immunosuppressive properties of MeV.

7. Perspectives

Many findings have enlarged our knowledge of the molecules and mechanisms involved in the establishment and/or the control of the immune response induced by MeV. Recently, the discovery of cell-free MeV N has led to propose new mechanisms by which MeV can simultaneously induce a strong and life-lasting anti-viral immune response and a subsequent profound immunosuppression. MeV N is the first viral nucleoprotein described for its ability to bind not only to the activatory receptor N-specific BCR, but also to the inhibitory FcγRII and NR receptors. As MeV N receptors are widely expressed amongst immune cells, MeV N can either up-regulate or inhibit the function of APCs (such as DCs), B and T cells. The characterization of NR will prove unvaluable for our understanding of the complex interplay between pathogens and the immune system. Several questions need yet to be addressed. What function(s) does MeV N trigger when associated with a single or multiple extracellular partners? Does the post-translational modification of N (i.e. phosphorylation) influence its immunologic properties? Likewise, are their any viral- or cell-derived partners naturally released with MeV N, which can induce specific conformational change(s) within N affecting its ability to interact with the immune system? Such conformational changes may indeed cause MeV N to bind to extracellular partners with different efficiency and to exhibit different specificities. These questions can only be answered once a precise relationship between the structure and the functions of this pleiotropic viral protein is established.

Acknowledgements

We particularly thank Drs S. Longhi and J.M. Bourhis for providing recombinant MeV N from bacteria. We are grateful for Drs C. Delprat and N. Etchart for helpful discussions. We also thank C. Bonnerot and S. Amigorena for providing IIA1.6 cells transfected with murine FcγRII, Shering Ploungh (France) for the L cell line and L cells expressing mFcγRIIb1, and David Y. Masson for the KB61 mAb. We are grateful to Dr G. Salles for providing Thomas, Me 275, and KMM1 cell lines. This work benefited from the technical facilities of the Centre Commun d'Imagerie de Laennec (CeCIL). We also thank C. Bella (Flow cytometry and Imaging/Microscopy platforms of IFR 128) for helpful technical assistance. The authors also express their gratitude to the autoimmunity and transplantation division staff at the Walter and Aliza Hall Institute (Australia) for their excellent suggestions and support.

The work performed by the authors and described in this review was supported in part by grants ANR (DG, ANR-MIME), INSERM, INCA-Canceropole 2004-2005 and ARC 3637. D.L. and P.O.V. were supported by a fellowship from ARC and MENRT, respectively. A.G. was supported by a fellowship from the ministry of higher education-Tishreen University-Syria.

References

[1] Amigorena, S., J. Salamero, J. Davoust, W. H. Fridman, and C. Bonnerot. 1992. Tyrosine-containing motif that transduces cell activation signals also determines internalization and antigen presentation via type III receptors for IgG. *Nature* 358:337-41.

[2] Arneborn, P., and G. Biberfeld. 1983. T-lymphocyte subpopulations in relation to immunosuppression in measles and varicella. *Infect. Immun.* 39:29-37.

[3] Atabani, S. F., A. A. Byrnes, A. Jaye, I. M. Kidd, A. F. Magnusen, H. Whittle, and C. L. Karp. 2001. Natural measles causes prolonged suppression of interleukin-12 production. *J. Infect. Dis.* 184:1-9.

[4] Auwaerter, P. G., H. Kaneshima, J. M. McCune, G. Wiegand, and D. E. Griffin. 1996. Measles virus infection of thymic epithelium in the SCID-hu mouse leads to thymocyte apoptosis. *J. Virol.* 70:3734-40.

[5] Auwaerter, P. G., P. A. Rota, W. R. Elkins, R. J. Adams, T. DeLozier, Y. Shi, W. J. Bellini, B. Murphy, and D. E. Griffin. 1999. Measles Virus Infection in Rhesus Macaques: Altered Immune Responses and Comparison of the Virulence of Six Different Virus Strains. *J. Infect. Dis.* 180:950–958.

[6] Banchereau, J., F. Briere, C. Caux, J. Davoust, S. Lebecque, Y. J. Liu, B. Pulendran, and K. Palucka. 2000. Immunobiology of dendritic cells. *Annu. Rev. Immunol.* 18:767-811.

[7] Bankamp, B., S. M. Horikami, P. D. Thompson, M. Huber, M. Billeter, and S. A. Moyer. 1996. Domains of the measles virus N protein required for binding to P protein and self-assembly. *Virology* 216:272-277.

[8] Beauverger, P., A. I. Cardoso, L. Daviet, R. Buckland, and T. F. Wild. 1996. Analysis of the contribution of CTL epitopes in the immunobiology of morbillivirus infection. *Virology* 219:133-9.

[9] Beauverger, P., J. Chadwick, R. Buckland, and T. F. Wild. 1994. Serotype-specific and canine distemper virus cross-reactive H-2Kk-restricted cytotoxic T lymphocyte epitopes in the measles virus nucleoprotein. *Virology* 203:172-7.

[10] Bech, V. 1959. Studies on the development of complement fixing antibodies in measles patients. Observations during a measles epidemic in Greenland. *J. Immunol.* 83:267-75.

[11] Bernasconi, N. L., N. Onai, and A. Lanzavecchia. 2003. A role for Toll-like receptors in acquired immunity: up-regulation of TLR9 by BCR triggering in naive B cells and constitutive expression in memory B cells. *Blood* 101:4500-4.

[12] Bernasconi, N. L., E. Traggiai, and A. Lanzavecchia. 2002. Maintenance of serological memory by polyclonal activation of human memory B cells. *Science* 298:2199-202.

[13] Bolland, S., R. N. Pearse, T. Kurosaki, and J. V. Ravetch. 1998. SHIP modulates immune receptor responses by regulating membrane association of Btk. *Immunity* 8:509-16.

[14] Boruchov, A. M., G. Heller, M. C. Veri, E. Bonvini, J. V. Ravetch, and J. W. Young. 2005. Activating and inhibitory IgG Fc receptors on human DCs mediate opposing functions. *J. Clin. Invest.* 115:2914-23.

[15] Bourhis, J., K. Johansson, V. Receveur-Bréchot, C. J. Oldfield, A. K. Dunker, B. Canard, and S. Longhi. 2004. The C-terminal domain of measles virus nucleoprotein belongs to the classof intrinsically disordered proteins that fold upon binding to their pohysiological partner. *Virus Research* 99:157-67.

[16] Bourhis, J. M., B. Canard, and S. Longhi. 2006. Structural disorder within the replicative complex of measles virus: functional implications. *Virology* 344:94-110.

[17] Bourhis, J. M., V. Receveur-Brechot, M. Oglesbee, X. Zhang, M. Buccellato, H. Darbon, B. Canard, S. Finet, and S. Longhi. 2005. The intrinsically disordered C-terminal domain of the measles virus nucleoprotein interacts with the C-terminal domain of the phosphoprotein via two distinct sites and remains predominantly unfolded. *Protein Sci.* 14:1975-92.

[18] Buckland, R., P. Giraudon, and F. Wild. 1989. Expression of measles virus nucleoprotein in Escherichia coli: use of deletion mutants to locate the antigenic sites. *J. Gen. Virol.* 70:435-41.

[19] Cardoso, A. I., P. Beauverger, D. Gerlier, T. F. Wild, and C. Rabourdin-Combe. 1995. Formaldehyde inactivation of measles virus abolishes CD46-dependent presentation of nucleoprotein to murine class I-restricted CTLs but not to class II-restricted helper T cells. *Virology* 212:255-8.

[20] Casali, P., G. P. Rice, and M. B. Oldstone. 1984. Viruses disrupt functions of human lymphocytes. Effects of measles virus and influenza virus on lymphocyte-mediated killing and antibody production. *J. Exp. Med.* 159:1322-37.

[21] Cassel, D. L., M. A. Keller, S. Surrey, E. Schwartz, A. D. Schreiber, E. F. Rappaport, and S. E. McKenzie. 1993. Differential expression of Fc gamma RIIA, Fc gamma RIIB and Fc gamma RIIC in hematopoietic cells: analysis of transcripts. *Mol. Immunol.* 30:451-60.

[22] Chino, F., H. Kodama, T. Ohkawa, and Y. Egashira. 1979. Alterations of the thymus and peripheral lymphoid tissues in fatal measles. A review of 14 autopsy cases. *Acta Pathol. Jpn.* 29:493-507.

[23] Coovadia, H. M., M. A. Parent, W. E. Loening, A. Wesley, B. Burgess, F. Hallett, P. Brain, J. Grace, J. Naidoo, P. M. Smythe, and G. H. Vos. 1974. An evaluation of factors associated with the depression of immunity in malnutrition and in measles. *Am. J. Clin. Nutr.* 27:665-9.

[24] Daeron, M., S. Latour, O. Malbec, E. Espinosa, P. Pina, S. Pasmans, and W. H. Fridman. 1995. The same tyrosine-based inhibition motif, in the intracytoplasmic domain of Fc gamma RIIB, regulates negatively BCR-, TCR-, and FcR-dependent cell activation. *Immunity* 3:635-46.

[25] Demotz, S., W. Ammerlaan, P. Fournier, C. P. Muller, and C. Barbey. 1998. Processing of the DRB1*1103-restricted measles virus nucleoprotein determinant 185-199 in the endosomal compartment. *Clin. Exp. Immunol.* 114:228-35.

[26] Devaux, P., D. Christiansen, S. Plumet, and D. Gerlier. 2004. Cell surface activation of the alternative complement pathway by the fusion protein of measles virus. *J. Gen. Virol.* 85:1665-1673.

[27] Dhanji, S., K. Tse, and H. S. Teh. 2005. The low affinity Fc receptor for IgG functions as an effective cytolytic receptor for self-specific CD8 T cells. *J. Immunol.* 174:1253-8.

[28] Diallo, A., T. Barrett, M. Barbron, G. Meyer, and P. C. Lefevre. 1994. Cloning of the nucleocapsid protein gene of peste-des-petits-ruminants virus: relationship to other morbilliviruses. *J. Gen. Virol.* 75:233-237.

[29] El Mubarak, H. S., S. A. Ibrahim, H. W. Vos, M. M. Mukhtar, O. A. Mustafa, T. F. Wild, A. D. Osterhaus, and R. L. De Swart. 2004. Measles virus protein-specific IgM, IgA, and IgG subclass responses during the acute and convalescent phase of infection. *J. Med. Virol.* 72:290-8.

[30] Erlenhoefer, C., W. J. Wurzer, S. Loffler, S. Schneider-Schaulies, V. ter Meulen, and J. Schneider-Schaulies. 2001. CD150 (SLAM) is a receptor for measles virus but is not involved in viral contact-mediated proliferation inhibition. *J. Virol.* 75:4499-505.

[31] Etchart, N., P. O. Desmoulins, K. Chemin, C. Maliszewski, B. Dubois, F. Wild, and D. Kaiserlian. 2001. Dendritic cells recruitment and in vivo priming of CD8+ CTL induced by a single topical or transepithelial immunization via the buccal mucosa with measles virus nucleoprotein. *J. Immunol.* 167:384-91.

[32] Forthal, D. N., G. Landucci, A. Habis, M. Zartarian, J. Katz, and J. G. Tilles. 1994. Measles virus-specific functional antibody responses and viremia during acute measles. *J. Infect. Dis*. 169:1377-80.

[33] Fournier, P., W. Ammerlaan, D. Ziegler, C. Giminez, C. Rabourdin-Combe, B. T. Fleckenstein, K. H. Wiesmuller, G. Jung, F. Schneider, and C. P. Muller. 1996. Differential activation of T cells by antibody-modulated processing of the flanking sequences of class II-restricted peptides. *Int. Immunol.* 8:1441-51.

[34] Fugier-Vivier, I., C. Servet-Delprat, P. Rivailler, M. C. Rissoan, Y. J. Liu, and C. Rabourdin-Combe. 1997. Measles virus suppresses cell-mediated immunity by interfering with the survival and functions of dendritic and T cells. *J. Exp. Med.* 186:813-23.

[35] Galama, J. M., J. Ubels-Postma, A. Vos, and C. J. Lucas. 1980. Measles virus inhibits acquisition of lymphocyte functions but not established effector functions. *Cell* Immunol 50:405-15.

[36] Gerlier, D., M. C. Trescol-Biemont, G. Varior-Krishnan, D. Naniche, I. Fugier-Vivier, and C. Rabourdin-Combe. 1994. Efficient major histocompatibility complex class II-restricted presentation of measles virus relies on hemagglutinin-mediated targeting to its cellular receptor human CD46 expressed by murine B cells. *J. Exp. Med.* 179:353-8.

[37] Gerlier, D., H. Valentin, D. Laine, C. Rabourdin-Combe, and C. Servet-Delprat. 2006. Subversion of the immune system by measles virus: a model for the intricate interplay between a virus and the immune sytem, p. 225-292. *In* P. Lachmman and M. B. Oldstone (ed.), *Microbial subvertion of host immunity*, vol. in press. Horizon Scientfic Press, Norfolk, UK.

[38] Giraudon, P., M. F. Jacquier, and T. F. Wild. 1988. Antigenic analysis of African measles virus field isolates: identification and localisation of one conserved and two variable epitope sites on the NP protein. *Virus Res.* 10:137-52.

[39] Giraudon, P., and T. F. Wild. 1981. Monoclonal antibodies against measles virus. *J. Gen. Virol.* 54:325-332.

[40] Godfraind, C., K. V. Holmes, and J. P. Coutelier. 1995. Thymus involution induced by mouse hepatitis virus A59 in BALB/c mice. *J. Virol.* 69:6541-7.

[41] Goldman, M. B., T. A. O'Bryan, D. J. Buckthal, L. M. Tetor, and J. N. Goldman. 1995. Suppression of measles virus expression by noncytolytic antibody in an immortalized macrophage cell line. *J. Virol*. 69:734-40.

[42] Good, R. A., and S. J. Zak. 1956. Disturbances in gamma globulin synthesis as"experiments of nature". *Pediatrics* 18:109-149.

[43] Graves, M., D. E. Griffin, R. T. Johnson, R. L. Hirsch, I. L. de Soriano, S. Roedenbeck, and A. Vaisberg. 1984. Development of antibody to measles virus polypeptides during complicated and uncomplicated measles virus infections. *J. Virol.* 49:409-412.

[44] Grazia Cappiello, M., F. S. Sutterwala, G. Trinchieri, D. M. Mosser, and X. Ma. 2001. Suppression of Il-12 transcription in macrophages following Fc gamma receptor ligation. *J. Immunol.* 166:4498-506.

[45] Griffin, D. E. 1995. Immune responses during measles virus infection. Curr Top Microbiol Immunol 191:117-34.

[46] Griffin, D. E. 2001. Measles virus, p. 1401-1441. *In* D. M. Knipe and P. M. Howley (ed.), *Fields Virology*, vol. 1. Lippincott-Raven, Philadelphia.
[47] Griffin, D. E., and B. J. Ward. 1993. Differential CD4 T cell activation in measles. *J. Infect. Dis.* 168:275-81.
[48] Griffin, D. E., B. J. Ward, and L. M. Esolen. 1994. Pathogenesis of measles virus infection: an hypothesis for altered immune responses. *J. Infect. Dis.* 170:S24-S31.
[49] Hahm, B., N. Arbour, D. Naniche, D. Homann, M. Manchester, and M. B. Oldstone. 2003. Measles virus infects and suppresses proliferation of T lymphocytes from transgenic mice bearing human signaling lymphocytic activation molecule. *J. Virol.* 77:3505-15.
[50] Hayashi, F., K. D. Smith, A. Ozinsky, T. R. Hawn, E. C. Yi, D. R. Goodlett, J. K. Eng, S. Akira, D. M. Underhill, and A. Aderem. 2001. The innate immune response to bacterial flagellin is mediated by Toll-like receptor 5. *Nature* 410:1099-103.
[51] Heyman, B. 2003. Feedback regulation by IgG antibodies. *Immunol. Lett.* 88:157-61.
[52] Hirsch, R. L., D. E. Griffin, R. T. Johnson, S. J. Cooper, I. Lindo de Soriano, S. Roedenbeck, and A. Vaisberg. 1984. Cellular immune responses during complicated and uncomplicated measles virus infections of man. *Clin. Immunol. Immunopath.* 31:1-12.
[53] Horejs-Hoeck, J., A. Hren, G. C. Mudde, and M. Woisetschlager. 2005. Inhibition of immunoglobulin E synthesis through Fc gammaRII (CD32) by a mechanism independent of B-cell receptor co-cross-linking. *Immunology* 115:407-15.
[54] Ilonen, J., M. J. Makela, B. Ziola, and A. A. Salmi. 1990. Cloning of human T cells specific for measles virus haemagglutinin and nucleocapsid. *Clin. Exp. Immunol.* 81:212-217.
[55] Isakov, N. 1997. ITIMs and ITAMs. The Yin and Yang of antigen and Fc receptor-linked signaling machinery. *Immunol. Res.* 16:85-100.
[56] Jacobson, S., R. P. Sekaly, C. L. Jacobson, H. F. McFarland, and E. O. Long. 1989. HLA class II-restricted presentation of cytoplasmic measles virus antigens to cytotoxic T cells. *J. Virol.* 63:1756-1762.
[57] Joffe, M. I., and A. R. Rabson. 1981. Defective helper factor (LMF) production in patients with acute measles infection. *Clin. Immunol. Immunopathol.* 20:215-23.
[58] Johansson, K., J. M. Bourhis, V. Campanacci, C. Cambillau, B. Canard, and S. Longhi. 2003. Crystal structure of the measles virus phosphoprotein domain responsible for the induced folding of the C-terminal domain of the nucleoprotein. *J. Biol. Chem.* 278:44567-73.
[59] Karlin, D., F. Ferron, B. Canard, and S. Longhi. 2003. Structural disorder and modular organization in Paramyxovirinae N and P. *J. Gen. Virol.* 84:3239-52.
[60] Karlin, D., S. Longhi, and B. Canard. 2002. Substitution of two residues in the measles virus nucleoprotein results in an impaired self-association. *Virology* 302:420-432.
[61] Karp, C. L., M. Wysocka, L. M. Wahl, J. M. Ahearn, P. J. Cuomo, B. Sherry, G. Trinchieri, and D. E. Griffin. 1996. Mechanism of suppression of cell-mediated immunity by measles virus. *Science* 273:228-31.

[62] Kerdiles, Y. M., B. Cherif, J. C. Marie, N. Tremillon, B. Blanquier, G. Libeau, A. Diallo, T. F. Wild, M. B. Villiers, and B. Horvat. 2006. Immunomodulatory properties of morbillivirus nucleoproteins. *Viral. Immunol.* 19:324-34.

[63] Kerdiles, Y. M., C. I. Sellin, J. Druelle, and B. Horvat. 2006. Immunosuppression caused by measles virus: role of viral proteins. *Rev. Med. Virol.* 16:49-63.

[64] Kobune, F., H. Takahashi, K. Terao, T. Ohkawa, Y. Ami, Y. Suzaki, N. Nagata, H. Sakata, K. Yamanouchi, and C. Kai. 1996. Nonhuman primate models of measles. *Lab. Anim. Sci.* 46:315-20.

[65] Kruisbeek, A. M. 1999. Introduction: regulation of T cell development by the thymic microenvironment. *Semin. Immunol.* 11:1-2.

[66] Laine, D., J. M. Bourhis, S. Longhi, M. Flacher, L. Cassard, B. Canard, C. Sautes-Fridman, C. Rabourdin-Combe, and H. Valentin. 2005. Measles virus nucleoprotein induces cell-proliferation arrest and apoptosis through NTAIL-NR and NCORE-FcgammaRIIB1 interactions, respectively. *J. Gen. Virol.* 86:1771-84.

[67] Laine, D., M. C. Trescol-Biemont, S. Longhi, G. Libeau, J. C. Marie, P. O. Vidalain, O. Azocar, A. Diallo, B. Canard, C. Rabourdin-Combe, and H. Valentin. 2003. Measles virus (MV) nucleoprotein binds to a novel cell surface receptor distinct from FcgammaRII via its C-terminal domain: role in MV-induced immunosuppression. *J. Virol.* 77:11332-46.

[68] Lanzavecchia, A., N. Bernasconi, E. Traggiai, C. R. Ruprecht, D. Corti, and F. Sallusto. 2006. Understanding and making use of human memory B cells. *Immunol. Rev.* 211:303-9.

[69] Le Borgne, M., N. Etchart, A. Goubier, S. A. Lira, J. C. Sirard, N. van Rooijen, C. Caux, S. Ait-Yahia, A. Vicari, D. Kaiserlian, and B. Dubois. 2006. Dendritic cells rapidly recruited into epithelial tissues via CCR6/CCL20 are responsible for CD8+ T cell crosspriming in vivo. *Immunity* 24:191-201.

[70] Longhi, S., V. Receveur-Brechot, D. Karlin, K. Johansson, H. Darbon, D. Bhella, R. Yeo, S. Finet, and B. Canard. 2003. The C-terminal domain of the measles virus nucleoprotein is intrinsically disordered and folds upon binding to the C-terminal moiety of the phosphoprotein. *J. Biol. Chem.* 278:18638-48.

[71] Malbec, O., J. P. Attal, W. H. Fridman, and M. Daeron. 2002. Negative regulation of mast cell proliferation by FcgammaRIIB. *Mol. Immunol.* 38:1295-9.

[72] Marie, J. C., J. Kehren, M. C. Trescol-Biemont, A. Evlashev, H. Valentin, T. Walzer, R. Tedone, B. Loveland, J. F. Nicolas, C. Rabourdin-Combe, and B. Horvat. 2001. Mechanism of measles virus-induced suppression of inflammatory immune responses. *Immunity* 14:69-79.

[73] Marie, J. C., F. Saltel, J. M. Escola, P. Jurdic, T. F. Wild, and B. Horvat. 2004. Cell Surface Delivery of the Measles Virus Nucleoprotein: a Viral Strategy To Induce Immunosuppression†. *J. Virol.* 78:11952–11961.

[74] Marttila, J., J. Ilonen, E. Norrby, and A. Salmi. 1999. Characterization of T cell epitopes in measles virus nucleoprotein. *J. Gen. Virol.* 80 (Pt 7):1609-15.

[75] McChesney, M. B., J. H. Kehrl, A. Valsamakis, A. S. Fauci, and M. B. Oldstone. 1987. Measles virus infection of B lymphocytes permits cellular activation but blocks progression through the cell cycle. *J. Virol.* 61:3441-7.

[76] Metes, D., L. K. Ernst, W. H. Chambers, A. Sulica, R. B. Herberman, and P. A. Morel. 1998. Expression of functional CD32 molecules on human NK cells is determined by an allelic polymorphism of the FcgammaRIIC gene. *Blood* 91:2369-80.

[77] Moench, T. R., D. E. Griffin, C. R. Obriecht, A. J. Vaisberg, and R. T. Johnson. 1988. Acute measles in patients with and without neurological involvement: distribution of measles virus antigen and RNA. *J. Infect. Dis.* 158:433-42.

[78] Moss, W. J., J. J. Ryon, M. Monze, and D. E. Griffin. 2002. Differential regulation of interleukin IL-4, IL-5, and IL-10 during measles in Zambian children. *J. Infect. Dis.* 186:879-87.

[79] Norrby, E., and Y. Gollmar. 1972. Appearance and persistence of antibodies against different virus components after regular measles infections. *Infect. Immun.* 6:240-7.

[80] Nozawa, Y., N. Ono, M. Abe, H. Sakuma, and H. Wakasa. 1994. An immunohistochemical study of Warthin-Finkeldey cells in measles. *Pathol. Int.* 44:442-7.

[81] Obeid, O. E., C. D. Partidos, C. R. Howard, and M. W. Steward. 1995. Protection against morbillivirus-induced encephalitis by immunization with a rationally designed synthetic peptide vaccine containing B- and T-cell epitopes from the fusion protein of measles virus. *J. Virol.* 69:1420-8.

[82] Oglesbee, M. J., H. Kenney, T. Kenney, and S. Krakowka. 1993. Enhanced production of morbillivirus gene-specific RNAs following induction of the cellular stress response in stable persistent infection. *Virology* 192:556-67.

[83] Okada, H., F. Kobune, T. A. Sato, T. Kohama, Y. Takeuchi, T. Abe, N. Takayama, T. Tsuchiya, and M. Tashiro. 2000. Extensive lymphopenia due to apoptosis of uninfected lymphocytes in acute measles patients. *Arch. Virol.* 145:905-20.

[84] Okada, H., T. A. Sato, A. Katayama, K. Higuchi, K. Shichijo, T. Tsuchiya, N. Takayama, Y. Takeuchi, T. Abe, N. Okabe, and M. Tashiro. 2001. Comparative analysis of host responses related to immunosuppression between measles patients and vaccine recipients with live attenuated measles vaccines. *Arch. Virol.* 146:859-74.

[85] Olszewska, W., J. Erume, J. Ripley, M. W. Steward, and C. D. Partidos. 2001. Immune responses and protection induced by mucosal and systemic immunisation with recombinant measles nucleoprotein in a mouse model of measles virus-induced encephalitis. *Arch. Virol.* 146:293-302.

[86] Ono, M., H. Okada, S. Bolland, S. Yanagi, T. Kurosaki, and J. V. Ravetch. 1997. Deletion of SHIP or SHP-1 reveals two distinct pathways for inhibitory signaling. *Cell* 90:293-301.

[87] Panum, P. L. 1847. Beobachtungen uber das Maserncontagious. *Virchows* 1:492-502.

[88] Pearse, R. N., T. Kawabe, S. Bolland, R. Guinamard, T. Kurosaki, and J. V. Ravetch. 1999. SHIP recruitment attenuates Fc gamma RIIB-induced B cell apoptosis. *Immunity* 10:753-60.

[89] Permar, S. R., S. A. Klumpp, K. G. Mansfield, A. A. Carville, D. A. Gorgone, M. A. Lifton, J. E. Schmitz, K. A. Reimann, F. P. Polack, D. E. Griffin, and N. L. Letvin. 2004. Limited contribution of humoral immunity to the clearance of measles viremia in rhesus monkeys. *J. Infect. Dis.* 190:998-1005.

[90] Permar, S. R., S. A. Klumpp, K. G. Mansfield, W. K. Kim, D. A. Gorgone, M. A. Lifton, K. C. Williams, J. E. Schmitz, K. A. Reimann, M. K. Axthelm, F. P. Polack, D. E. Griffin, and N. L. Letvin. 2003. Role of CD8 Lymphocytes in Control and Clearance of Measles Virus Infection of Rhesus Monkeys. *J. Virol.* 77:4396–4400.

[91] Ravanel, K., C. Castelle, T. Defrance, T. F. Wild, D. Charron, V. Lotteau, and C. Rabourdin-Combe. 1997. Measles virus nucleocapsid protein binds to FcgammaRII and inhibits human B cell antibody production. *J. Exp. Med.* 186:269-278.

[92] Ravetch, J. V., and S. Bolland. 2001. IgG Fc receptors. Annu Rev Immunol 19:275-90.

[93] Ray, J., and R. S. Fujinami. 1987. Characterization of in vitro transcription and transcriptional products of measles virus. *J. Virol.* 61:3381-7.

[94] Reth, M., and J. Wienands. 1997. Initiation and processing of signals from the B cell antigen receptor. *Annu. Rev. Immunol.* 15:453-79.

[95] Ryon, J. J., W. J. Moss, M. Monze, and D. E. Griffin. 2002. Functional and phenotypic changes in circulating lymphocytes from hospitalized zambian children with measles. *Clin. Diagn. Lab. Immunol.* 9:994-1003.

[96] Sakaguchi, M., Y. Yoshikawa, K. Yamanouchi, T. Sata, K. Nagashima, and K. Takeda. 1986. Growth of measles virus in epithelial and lymphoid tissues of cynomolgus monkeys. *Microbiol. Immunol.* 30:1067-73.

[97] Sandilands, G. P., S. A. MacPherson, E. R. Burnett, A. J. Russell, I. Downie, and R. N. MacSween. 1997. Differential expression of CD32 isoforms following alloactivation of human T cells. *Immunology* 91:204-11.

[98] Scheller, C., and C. Jassoy. 2001. Syncytium formation amplifies apoptotic signals: a new view on apoptosis in HIV infection in vitro. *Virology* 282:48-55.

[99] Schnorr, J. J., M. Seufert, J. Schlender, J. Borst, I. C. Johnston, V. ter Meulen, and S. Schneider-Schaulies. 1997. Cell cycle arrest rather than apoptosis is associated with measles virus contact-mediated immunosuppression in vitro. *J. Gen. Virol.* 78:3217-26.

[100] Servet-Delprat, C., P. O. Vidalain, H. Bausinger, S. Manie, F. Le Deist, O. Azocar, D. Hanau, A. Fischer, and C. Rabourdin-Combe. 2000. Measles virus induces abnormal differentiation of CD40 ligand-activated human dendritic cells. *J. Immunol.* 164:1753-60.

[101] Servet-Delprat, C., P. O. Vidalain, H. Valentin, and C. Rabourdin-Combe. 2003. Measles virus and dendritic cell functions: how specific response cohabits with immunosuppression. *Curr. Top Microbiol. Immunol.* 276:103-23.

[102] Shimizu, T., N. Fukushima, Y. Ebihara, T. Sata, and Y. Aoyama. 1987. Giant cell pneumonia in Letterer-Siwe disease. *Acta Pathol. Jpn.* 37:493-501.

[103] Sissons, J. G. P., R. D. Schreiber, L. H. Perrin, N. R. Cooper, H. J. Muller-Eberhard, and M. B. A. Oldstone. 1979. Lysis of measles virus-infected cells by the purified cytolytic alternative complement pathway and antibody. *J. Exp. Med.* 150:415-454.

[104] Slifka, M. K., and R. Ahmed. 1998. Long-lived plasma cells: a mechanism for maintaining persistent antibody production. *Curr. Opin. Immunol.* 10:252-8.

[105] Speziani, C., D. Laine, C. Servet-Delprat, H. Valentin, and C. Rabourdin-Combe. 2004. Measles virus and immunosuppression. *Med. Mal. Infect.* 34:S2-S6.

[106] Tajima, M., and S. Kudow. 1976. Morphology of the Warthin-Finkeldey giant cells in monkeys with experimentally induced measles. *Acta Pathol. Jpn.* 26:367-80.

[107] Tew, J. G., J. Wu, D. Qin, S. Helm, G. F. Burton, and A. K. Szakal. 1997. Follicular dendritic cells and presentation of antigen and costimulatory signals to B cells. *Immunol. Rev.* 156:39-52.

[108] Tsubata, T. 1999. Co-receptors on B lymphocytes. *Curr. Opin. Immunol.* 11:249-55.

[109] Ueda, K., T. Imamura, T. Kawaguchi, H. Tamari, and T. Kambara. 1982. Subacute sclerosing panencephalitis. An autopsy case with impaired cellular immunity. *Acta Pathol. Jpn.* 32:1103-10.

[110] Uher, F., and H. B. Dickler. 1986. Cooperativity between B lymphocyte membrane molecules: independent ligand occupancy and cross-linking of antigen receptors and Fc gamma receptors down-regulates B lymphocyte function. *J. Immunol.* 137:3124-9.

[111] van Binnendijk, R. S., M. C. Poelen, P. de Vries, H. O. Voorma, A. D. Osterhaus, and F. G. Uytdehaag. 1989. Measles virus-specific human T cell clones. Characterization of specificity and function of CD4+ helper/cytotoxic and CD8+ cytotoxic T cell clones. *J. Immunol.* 142:2847-2854.

[112] Vasconcelos, D. Y., X. H. Cai, and M. J. Oglesbee. 1998. Constitutive overexpression of the major inducible 70 kDa heat shock protein mediates large plaque formation by measles virus. *J. Gen. Virol.* 79:2239-47.

[113] Vidalain, P. O., O. Azocar, B. Lamouille, A. Astier, C. Rabourdin-Combe, and C. Servet-Delprat. 2000. Measles virus induces functional TRAIL production by human dendritic cells. *J. Virol.* 74:556-9.

[114] Vidalain, P. O., O. Azocar, C. Rabourdin-Combe, and C. Servet-Delprat. 2001. Measle virus-infected dendritic cells develop immunosuppressive and cytotoxic activities. *Immunobiology* 204:629-38.

[115] Vincent, S., D. Spehner, S. Manie, R. Delorme, R. Drillien, and D. Gerlier. 1999. Inefficient measles virus budding in murine L.CD46 fibroblasts. Virology 265:185-95.

[116] Vivier, E., and M. Daeron. 1997. Immunoreceptor tyrosine-based inhibition motifs. *Immunol. Today* 18:286-91.

[117] Von Pirquet, C. 1908. Das verhalten der kutanen tuberculinreaktion wahrend der masern. *Deutche Midizinische Wochenschrift* 34:1297.

[118] Wang, M., J. E. Libbey, I. Tsunoda, and R. S. Fujinami. 2003. Modulation of immune system function by measles virus infection. II. Infection of B cells leads to the production of a soluble factor that arrests uninfected B cells in G0/G1. *Viral. Immunol.* 16:45-55.

[119] Ward, B. J., and D. E. Griffin. 1993. Changes in cytokine production after measles virus vaccination: predominant production of IL-4 suggests induction of a Th2 response. *Clin. Immunol. Immunopathol.* 67:171-7.

[120] Warmerdam, P. A., N. M. Nabben, S. A. van de Graaf, J. G. van de Winkel, and P. J. Capel. 1993. The human low affinity immunoglobulin G Fc receptor IIC gene is a result of an unequal crossover event. *J. Biol. Chem.* 268:7346-9.

[121] Wernersson, S., M. C. Karlsson, J. Dahlstrom, R. Mattsson, J. S. Verbeek, and B. Heyman. 1999. IgG-mediated enhancement of antibody responses is low in Fc receptor gamma chain-deficient mice and increased in Fc gamma RII-deficient mice. *J. Immunol.* 163:618-22.

[122] White, R. G., and J. F. Boyd. 1973. The effect of measles on the thymus and other lymphoid tissues. *Clin. Exp. Immunol.* 13:343-57.

[123] Whittle, H. C., A. Bradley-Moore, A. Fleming, and B. M. Greenwood. 1973. Effects of measles on the immune response of Nigerian children. *Arch Dis. Child* 48:753-6.

[124] Whittle, H. C., J. Dossetor, A. Oduloju, A. D. Bryceson, and B. M. Greenwood. 1978. Cell-mediated immunity during natural measles infection. *J. Clin. Invest.* 62:678-84.

[125] Yamanouchi, K., F. Chino, F. Kobune, H. Kodama, and T. Tsuruhara. 1973. Growth of measles virus in the lymphoid tissues of monkeys. *J. Infect. Dis.* 128:795-9.

[126] Yanagi, Y., B. A. Cubitt, and M. B. Oldstone. 1992. Measles virus inhibits mitogen-induced T cell proliferation but does not directly perturb the T cell activation process inside the cell. *Virology* 187:280-9.

[127] Zhang, X., J. M. Bourhis, S. Longhi, T. Carsillo, M. Buccellato, B. Morin, B. Canard, and M. Oglesbee. 2005. Hsp72 recognizes a P binding motif in the measles virus N protein C-terminus. *Virology* 337:162-74.

[128] Zhang, X., C. Glendening, H. Linke, C. L. Parks, C. Brooks, S. A. Udem, and M. Oglesbee. 2002. Identification and characterization of a regulatory domain on the carboxyl terminus of the measles virus nucleocapsid protein. *J. Virol.* 76:8737-8746.

Index

A

B

C

D

E

I

N

O

P

R

S

T

U

V

W

X

Y